Ancient Trees in the Landscape

Norfolk's arboreal heritage

Ancient Trees in the Landscape

Norfolk's arboreal heritage

by

Gerry Barnes and Tom Williamson

Windgather Press
is an imprint of
Oxbow Books

ISBN 978-1-905119-39-4

A CIP record for this book is available from the British Library

This book is available direct from

Oxbow Books, Oxford, UK
(Phone: 01865-241249; Fax: 01865-794449)

and

The David Brown Book Company
PO Box 511, Oakville, CT 06779, USA
(Phone: 860-945-9329; Fax: 860-945-9468)

or from our website

www.oxbowbooks.com

Printed in Great Britain by
Information Press, Eynsham, Oxfordshire

Contents

Abbreviations

BAR	British Archaeological Report
ESRO	East Suffolk Record Office (Ipswich)
NRO	Norfolk Record Office
NHER	Norfolk Historic Environment Record
TNA: PRO	The National Archives: Public Records Office, Kew
WSRO	West Suffolk Record Office (Bury St Edmunds)

Acknowledgements

We would like to thank the very large number of volunteers, parish tree wardens, friends and students who undertook the basic survey work upon which this study is based. Particular thanks must go to Brian Abbs, Gilbert Addison, Penny Andrews, John Arnott, Simon Aylmer, John Barber, Natalie Barnes, Clare Buxton, Carol Carpenter, Rod Chapman, Tony Codling, Nick Coleman, Chris Cook, Tom Cook, Henry Cordeaux, Chris Cowell, Peter Croot, Patsy Dallas, Richard Daplyn, J. T. Debbage, John Ebbage, the Rev. and Mrs B. Faulkener, John Fleetwood, Margaret Freeman, Mary Ghullan, D. F. Godwin, Bill Green, Jon Gregory, Ruth Hadman, Carol Haines, Ruth Hall, Sarah Harrison, Rory Hart, Peter Lambley, M. Lathan, Phil Lazaretti, Barry Leeson, Keith Lovett, Richard MacMullen, Dr A. J. Martin, Jamie Matthews, Colin McDonald, Jo Parmenter, Chris Roberts, Chris Shelton, Annie Sommazzi, Alan Spinks, Sarah Spooner, John Stockdale, Mike Tinsley, Henry Walker, Lucy Whittle, Nicola Whyte, Eric Wilkinson, Anne Wood and Andrew Woodhead. We would also like to thank the Heritage Lottery Fund for providing generous support for the survey.

Thanks are also due to innumerable landowners who allowed access to their properties; to the Norfolk Wildlife Trust; and to the staff of the Norfolk Record Office. Friends and colleagues at the University of East Anglia and in Norfolk County Council, and in Norfolk more generally, have also provided much advice, information or inspiration, especially Gary Battell, Nick Coleman, David Cubitt, Clare Dobbing, Jon Gregory, Sarah Harrison, Robert Liddiard, Annie Sommazzi, Sarah Spooner, Sue Thornett, John White, Jessica Williamson and Matt Williamson. Above all, we would like to thank Patsy Dallas, for all her research into orchards, wood pastures and much else, and for invaluable advice and ideas.

The photographs are by the authors, except 1, 5, 12, 15, 19 and 31, which are by Patsy Dallas and 55 and 56, which are by Graeme Cresswell. Figure 11 is reproduced courtesy of the Norfolk and Norwich Millennium Library and Mrs Flowerdew; Figures 25, 52 and 61 by permission of Norfolk Record Office. The maps and diagrams are by Jon Gregory, except for 39, which is by Andrew MacNair; and 40, 41, 66 and 68, which are by the authors.

List of Figures

CHAPTER ONE

Ancient Trees in the Landscape

Introduction

This book is about the trees which grow in one particular English county, Norfolk. It is largely – although not exclusively – about the larger, older specimens, and about those which have been managed in the past in 'traditional' ways. Yet although it deals in detail with only one, relatively restricted, area of the country, we believe that the work presented here has a wider relevance. While each region of Britain has its own, distinctive, arboricultural history the approaches and methodology adopted in this book can be applied elsewhere, and some of our general conclusions may well hold true for other areas, especially in the south and east of England. This is not, therefore, simply a book about old trees in Norfolk. It is about some of the ways in which we can study old trees.

Many people have written about the history of trees in the British landscape: historical ecologists such as the eminent Oliver Rackham; historical geographers such as Charles Watkins and Stephen Daniels; and landscape historians, especially Richard Muir (Rackham 1976 and 1986a; Petit and Watkins 2003; Muir 2005; Daniels 1988). The Tree Council has done much to draw the public's attention to the importance of our 'heritage trees', in part through the work of the Ancient Tree Forum (Fay 2002). It has sponsored a national volunteer survey (the 'Ancient Tree Hunt') (Thomas 2007) and published numerous articles on the subject in *Tree News*, as well as supporting the publication of Stokes and Rodger's informative and lavishly produced *The Heritage Trees of Britain and Northern Ireland* (Stokes and Rodgers 2004).

This book differs from most of those earlier studies in a number of respects. In particular, we are concerned above all with the landscape context of trees: with the questions of how, and why, particular kinds of old tree are found in particular places, what this has contributed to the character of the landscape, and what it can tell us about the ways in which trees were managed and regarded in the past – what trees *meant* to people in earlier periods. It is thus based on the results of a number of detailed systematic surveys and research projects which have been carried out over the years at the University of East Anglia, studies which – precisely because they have focused on one limited (if geographically diverse) region – allow us to tease out some of the complex links between the trees we see today and wider aspects of human and natural history. These links are perhaps less obvious when the subject is considered on a wider spatial canvas, or when trees are studied in isolation from the surrounding landscape. John White, in the title of one of his important articles on veteran trees, posed the question: 'What is a veteran tree and where are they all?' (White 1997). The research presented here attempted to answer a related but slightly different question: 'Where are all the veteran trees and what are they doing there?'

The most important of the projects whose results are presented in this book was a systematic survey, generously supported by the Heritage Lottery Fund and carried out over a number of years by volunteers, students and the authors themselves, of old trees in the county: broadly defined as all trees which have been pollarded in the past (Figure 1), have a circumference in excess of 4m, or which are noticeably old examples of their particular species. Details of over 5000 such trees were recorded and, while the survey did not aim to discover *every* example in Norfolk, coverage was fairly even in spatial terms, and the results seem to provide a good indication of the ways in which the age and character of our older trees varies across the county. The results of a number of smaller surveys, carried out by individuals and groups, were also made available to us. These included information collected by the Farming and Wildlife Advisory Group in connection with Higher Level Stewardship schemes, and by members of Norfolk County Council's countryside team as part of their work; surveys made for the various District Councils in Norfolk of veteran trees in the areas under their jurisdiction; an important survey of the trees found in the county's churchyards, carried out under the aegis of the Norfolk Wildlife Trust; and a detailed record of all the trees on the National Trust's Blickling and

FIGURE 1. The distinctive shape of this pollarded oak near Hethersett is particularly clear in the winter snow.

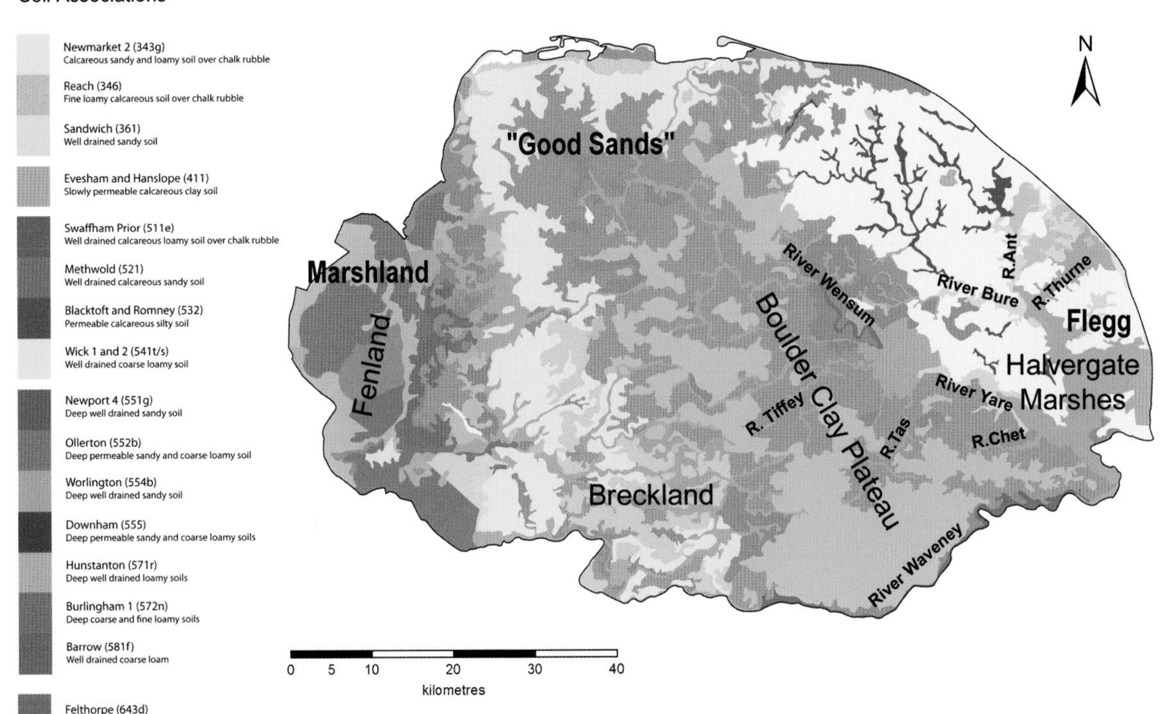

Figure 2. Soils and soil districts of Norfolk referred to in the text.

Felbrigg estates. In addition, during the late 1980s and early 1990s a number of tree surveys were undertaken as part of a wider study of the history of designed landscapes – parks and gardens – in the county, the full results of which have been published elsewhere (Williamson 1998). Although recording the character of the surviving planting formed only a small part of this project – and in only a limited number of cases were park and garden trees recorded in systematic detail – this survey also provided much information about changing fashions in planting and management. The current volume also draws on recent research into the history of orchards in the county, carried out largely by Patsy Dallas, and into the 'contorted pine rows' which form so distinctive a feature of the landscape of the sandy district of south-west Norfolk (and north-west Suffolk) known as Breckland. Lastly, we include here some discussion of the preliminary results of a programme of research, likely to continue for several more years, into the history and archaeology of ancient woodland in Norfolk.

Norfolk is a particularly good area in which to undertake a systematic study of old trees. In spite of some descriptions of the county's landscape as flat, dull and uninteresting, Norfolk is in fact very varied not only topographically but also in terms of its geology and soils. The only extensive areas of truly level terrain are to be found in the far west of the county – the drained wetlands of Fenland and Marshland – and in the east – the Halvergate Marshes (Figure 2) – although some parts of the boulder-clay plateau running through the centre and south of the county are also flat over several square kilometres of ground.

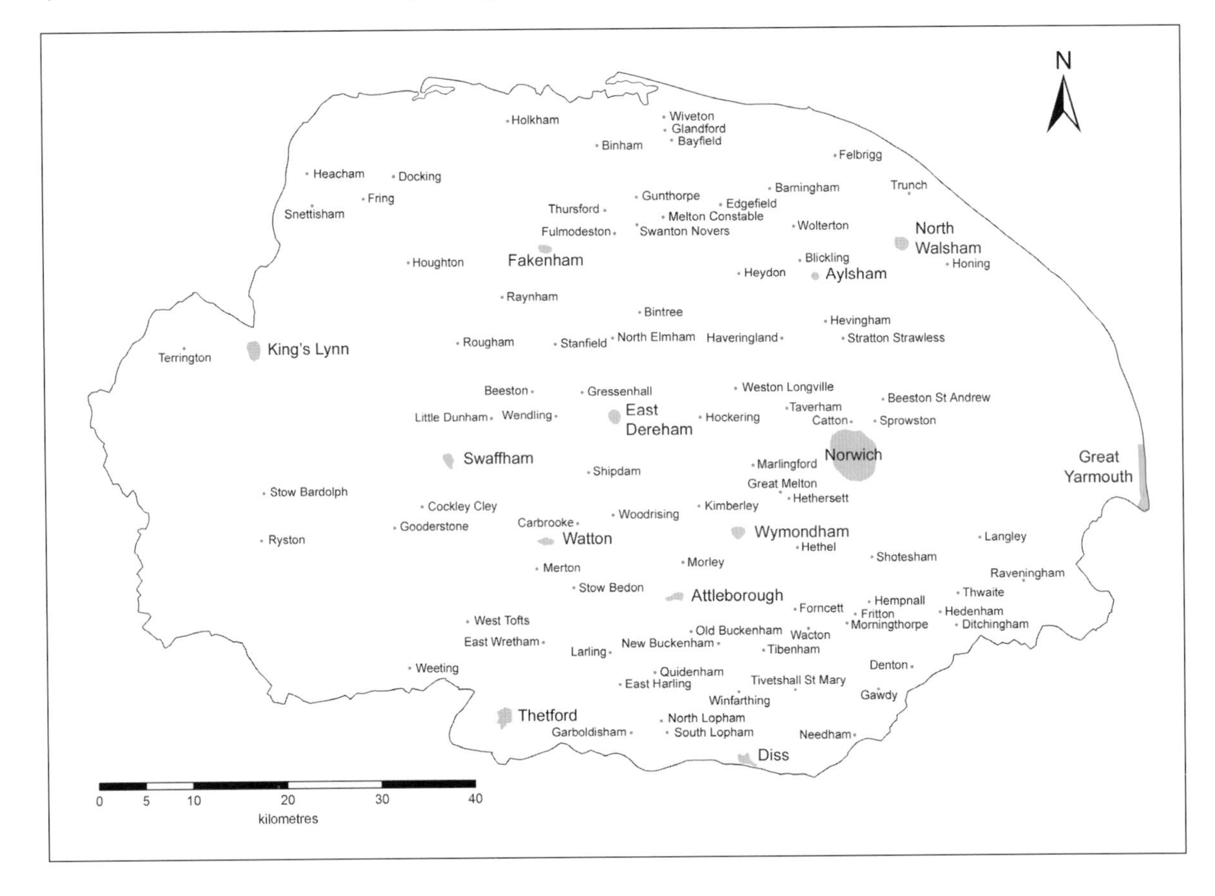

FIGURE 3. Some of the principal places in Norfolk referred to in the text.

Elsewhere, the land is either gently rolling or – in parts of the north and west especially – positively hilly. The soils are likewise diverse, ranging from silts and peat on the lower ground, through acid sands (most notably in Breckland) and damp boulder clays, to the fertile loams of Flegg and the north-east. But, as we shall see, it was less the intrinsic character of soils and topography, so much as the human landscapes of fields, commons, parks and settlements which these engendered, that have had a determining influence on our legacy of old trees.

Types of tree

Trees can be managed in a number of ways. They can be grown as *maidens* or *standards* – that is, left to grow in a natural way or with only minimal pruning. Such trees might be primarily ornamental specimens, in gardens or pleasure grounds or scattered across the open turf of a park; or they might be grown primarily for *timber* (large pieces employed in the construction of buildings and ships), in which case they were usually, in the past, felled when mature, in the case of oaks at between 80 and 100 years of age (Rackham 1986a, 65–7). Timber might be widely dispersed across the landscape, in hedges, parks and pastures; or it might be grouped into woods and plantations. Traditional woods were mainly, however, managed as *coppice-with-standards*: that is, the majority of trees and shrubs were coppiced, being repeatedly cut down to a *stool* at or near ground level on a rotation of eight to fifteen years, from which they rapidly regenerated to produce *wood* in the form of straight poles of the right size and shape for fencing, firewood, tool handles, minor structural elements in buildings and a host of other domestic uses (Rackham 1976; 1986b). Standard trees were also grown in such woods but they were usually quite widely spaced, at least in the period before the nineteenth century, for the shade cast by their canopy tended to inhibit the growth of the coppiced understorey beneath (Barnes 2003). Livestock were rigorously excluded from woods, or allowed in them only at certain times or under close supervision, because they would browse off the regenerating coppices and suppress their regrowth, ultimately destroying them. In other situations, where stock could not be so excluded, wood was produced from *pollards*. These were, in effect, aerial coppices, raised on a trunk, or *bolling*, at a height of around 2–3m, and thus out of the reach of browsing animals. Pollards could be grown in hedges or thinly scattered across pasture fields, but they might also be grouped into areas of *wood-pasture*, in which animals were regularly grazed but wood (and timber) were also produced. As we shall see, the majority of trees more than two centuries old in the Norfolk countryside, as elsewhere in England, are former pollards. Most have not been cropped for many decades – usually for more than a century – but they can still in most cases be easily recognised. A short (*c.*2–3m) trunk is topped by an area of damaged, calloused bark, from which the main branches of the tree all rise (Figure 4). Pollards may also have been used to provide browse or fodder for livestock, and this may have been the main function of *shreds*. These were trees from which the side branches were systematically removed to leave a small tuft at the top, a form of management probably adopted to encourage the development of twiggy, epicormic growth down the stem of the tree. Shredding seems to have been common in East Anglia before the seventeenth century, but declined rapidly thereafter. The documentary

FIGURE 4. An old oak pollard in the park at Heydon in north Norfolk.

references available to us make it clear that pollarding was usually carried out in the winter months, when leaves were off the trees; but if trees were being managed as a source of browse it presumably took place during the summer.

Our interest is primarily with old *trees*, but this raises some tricky questions of definition. Many common tree species in Norfolk, as elsewhere, were traditionally managed in such a way that they always remained, in effect, as low shrubs or bushes. Ash, for example, is relatively rare as an ancient tree, but many thousands of ancient stools, kept low for centuries (and often to this day) by regular coppicing, exist within the understorey of traditional woods or as part of hedges. We discuss some examples of ancient coppices and similar features in the course of this book, but our main focus is firmly on trees, whether maidens or old pollards: that is, on specimens with a fully formed trunk or bolling.

Because our interest is in both ancient trees and trees managed in traditional ways, the surveyors whose data provides the basis for much that is written here were, as already intimated, asked to record three main kinds of tree. Firstly, they noted all large specimens, roughly defined as any tree with a girth (or circumference), measured at waist height, of 4m or more. Secondly, they were asked to record all trees, regardless of size, which appeared to have been managed in the past by pollarding or shredding – what we term throughout this book 'traditionally managed' trees. Lastly, they were asked to note trees which, while smaller in girth than 4m, could nevertheless be regarded as of 'veteran' status.

FIGURE 5. A veteran elder in a hedge at Wymondham in south Norfolk. Trees do not have to be ancient to be of 'veteran' status: veterans are simply examples which are old for their particular species.

Veteran trees and biodiversity

We have already had cause to use the term 'veteran', which perhaps requires some explanation. It was coined and is principally employed not by landscape historians but by natural historians and arboriculturalists. In essence, a veteran tree is one which is not necessarily very ancient in terms of years, but which is an old example of its particular species (White 1997; 1998). While a veteran oak will thus be several centuries old, a silver birch will seldom live for more than a century and a half, making an example a hundred years old a 'veteran' specimen. A veteran elder might be younger still, perhaps no more than forty or fifty years of age (Figure 5). Whatever their species, however, veteran trees are of crucial importance because they provide a range of habitats, not otherwise generally

FIGURE 6. Veteran trees usually have numerous cavities and hollows which provide homes for owls, bats and a range of other vertebrates.

available, for a wide variety of plants and animals. They contain rotten wood, cavities, cracks and crevices which are home to rare lichens, fungi and beetles, and even to larger fauna, such as owls and bats (Briggs 1998; Kirby and Drake 1993; Orange 1994; Read and Frater 1999; Read 2000; Rose 1991) (Figure 6). They are particularly important because the natural environment of England, before human intervention, included substantial areas of wooded ground within which numerous trees grew old, died, fell and rotted, in contrast to their usual fate in later centuries, when they were often felled at economic maturity or cleared away and tidied up if brought down by natural causes.

Interest in veteran trees has grown steadily from the last decades of the twentieth century, fuelled in part by the books of writers such as Thomas Pakenham but more generally by an awareness of the importance of such trees as habitats (e.g. Morton 1998; Muir 2005; Pakenham 1996; Stokes and Rodgers 2004). English Nature's Veteran Tree Initiative was launched in 1996 and, while this came to an end in 2000, interest in the subject was maintained though the 'Lowland wood-pasture and parkland habitat action plan'. As a consequence of these initiatives the importance of ancient trees for biodiversity is now much more widely recognised by landowners, land agents and farmers, although poor management decisions about them are still sometimes made (as detailed in Green 2001, 166, and Green 2002).

A number of organisms are peculiarly associated with old trees, including a range of fungi which live off and destroy the old dead wood, degrading it in a variety of ways at cellular level to produce different kinds of decay. These can broadly be divided into white rot, in which the cellulose, lignin and hemicellulose in the wood is broken down; red and brown rot, in which the lignin remains intact but the cellulose is degraded (leading to the eventual formation of thick humus wood mould); and soft rot, in which cellulose, hemicellulose and lignin are all broken down, together with the cell walls. All these different forms of rot themselves provide habitats for a wide range of invertebrates, many extremely rare. At the same time, the fungal mycelia and fruiting bodies provide food (and often a habitat) for invertebrates and – by eventually leading to the hollowing of the trunk – for a number of vertebrates.

Invertebrates associated with ancient trees generally exhibit poor mobility: some are poor flyers, others are completely wingless. They are also, with a few notable exceptions, poor at breaking down cellulose and lignin, the main components of wood, and thus rely on the fungi (and other micro-organisms) to break it down for them. It has been estimated that there are no fewer than 1700 different species of invertebrates in Britain which depend on decayed wood for at least some of their life cycle (Alexander 1999, 109). The dependence of such saproxylic species can be direct (in that they feed on the fungi which decomposes the wood) or indirect (in that they predate on the former). Invertebrates also depend on such things as the loose bark and water-filled pockets which are characteristic of ancient trees. Large numbers of saproxylic species have been recorded from veteran trees in Norfolk, among which are twenty-two species of beetle, including extremely rare varieties such as *Agathidium confusum* (Collier 2001).

Veteran trees are also host to a range of lichens, many of which are, once again, largely restricted to this habitat and in particular to old trees growing in wood-pastures and parkland, where open glades provide the mixture of light and shelter which they need to thrive. The decayed structure of ancient trees, featuring numerous cavities, irregularities, areas of peeling bark and broken boughs, as well as the 'bark fluxes' caused by fungal or bacterial action, also provide an ideal range of niches for these often demanding organisms (Kirby and Drake 1993; Coleman 2002). Among the rare or unusual lichens found on veteran trees in Norfolk are *Lecanactis lyncea* and *L. premnea*. Some true plants – mosses such as *Zygodon forsteri*, liverworts and ferns – also grow on ancient trees, exploiting the fact that the irregularities in the bark surface provide shelter and permanently damp areas, while the often open canopy, punctuated by decayed branches, allows light to reach them. Snails and slugs such as the tree snail *Balera perversa*, feeding on both fungi and various mosses and ferns, also flourish, as do larger fauna, especially bats: of the fourteen species of bat found in Britain, twelve rely on trees, especially old trees, for shelter and/or food. Birds, especially owls such as the barn owl, nest in the numerous holes created as old trees decay. Mammals such as the stoat, weasel and dormouse also find shelter in ancient trees, while reptiles such as the great crested newt (*Triturus cristatus*) make their homes in cavities beneath loose bark and feed off the insects living there.

In short, veteran trees are host to a wide variety of plants and animals, reflecting both the extraordinary range of niches and habitats which they provide and the fact that, as has already been noted, much of our indigenous wildlife is adapted to an environment in which – as was presumably the case in the British 'wildwood' before the clearances, management and general tidiness wrought by man – large amounts of rotting wood were present. A single veteran tree might today be associated with a large number of organisms which, should that tree be removed, would find it very difficult to find another home, given that the nearest similar tree might be several kilometres away. It can constitute, in effect, a nature reserve in its own right. It is thus not surprising that some writers have argued that even individual trees should, in certain cases, be designated as Sites of Special Scientific Interest (Green 2001). Certainly, concentrations of old open-grown trees, in park-like or wood-pasture landscapes, are associated with particularly high levels of biodiversity (Green 2010).

Veteran trees thus have a vital role in maintaining biodiversity. But, in addition to all this, very old trees have a historic and cultural importance. Old trees, and especially those growing in farmland, are a crucial component of our historic landscapes and, perhaps more than any feature except vernacular buildings, help to define the essential character of particular areas and regions, providing them with much of their distinctive appearance (Stokes and Hand 2004). They also form a direct, living link with our ancestors, in some cases stretching back to medieval times. In a rapidly changing world they are thus symbols of stability and continuity, characteristics that they share with other features of the cultural landscape such as vernacular buildings and parish churches. The crucial difference, however, difficult for even the most objective of researchers to ignore, is that ancient trees are living organisms. As we shall see, interest in and respect for old trees is not a recent development: it goes back centuries. The study of old trees can also cast important new light on social and economic history, revealing much about the character of rural communities in the past and their attitudes to the world around them. In short, old trees are as much the concern of the historian as they are of the natural scientist.

But there are other reasons why we need to pay particular attention to old trees and their histories. Combined with other sources of information they can tell us much about the wider history of the countryside, and thus inform our decisions about how it should be managed in the future. The countryside of Norfolk, for example, is today dominated by hedgerow oaks and these are, quite understandably, often the main or only tree planted when hedges are re-established by landowners, usually as part of government-funded agro-environment schemes. But was oak always the dominant tree in the past, or is the present situation the consequence of relatively recent changes in the landscape? In a similar way, current projects to extend the area of heathland – in East Anglia as elsewhere in lowland England – are underpinned by the assumption that heaths were always open, treeless environments. But, as we shall see, an examination of ancient trees in the county casts some doubt on this assumption. Historical and archaeological approaches to trees can thus help natural scientists understand the character of the legacy of essentially semi-natural habitats which they study and manage. And in the pages that follow we treat trees, in some ways, as archaeological features. We are thus concerned with the morphology of individual specimens – with what their size and present form can tell us about their age and about how they were managed in the past. But we are perhaps more interested in looking at old trees, or particular species of old tree, *en masse*, in terms of their landscape context, because we can learn much by examining, for example, the kinds of tree that are associated with field boundaries of particular types, or how individual specimens in parks or pastures relate to the earthwork traces of past settlements or field systems. And, as with all archaeological enquiries, we need to understand, above all, the kinds of things that have shaped the character of the archaeological record: the processes of taphonomy. We need to understand something about how trees are lost from the landscape – often in selective, patterned ways – if we are to understand the meaning of the distribution and associations of those that survive today.

How many old trees are there, and where are they found?

Details of over 5500 veteran, ancient and traditionally managed trees have been recorded in Norfolk (Figure 7). The first questions that we need to ask are how this figure compares with the total number of such trees in the county and – more importantly – how representative this sample is in terms of location, species and age. Large surveys carried out by volunteers always carry the risk of uneven coverage. Keen and able individuals, willing to give up their free time, are themselves unevenly distributed. On the other hand, the authors have made every effort to investigate apparent gaps in the distribution of recorded trees, investigations which have served to confirm the overall pattern derived from volunteer observations. Systematic investigation of apparent *lacunae* has tended to maintain the apparent contrasts between areas in which old trees are

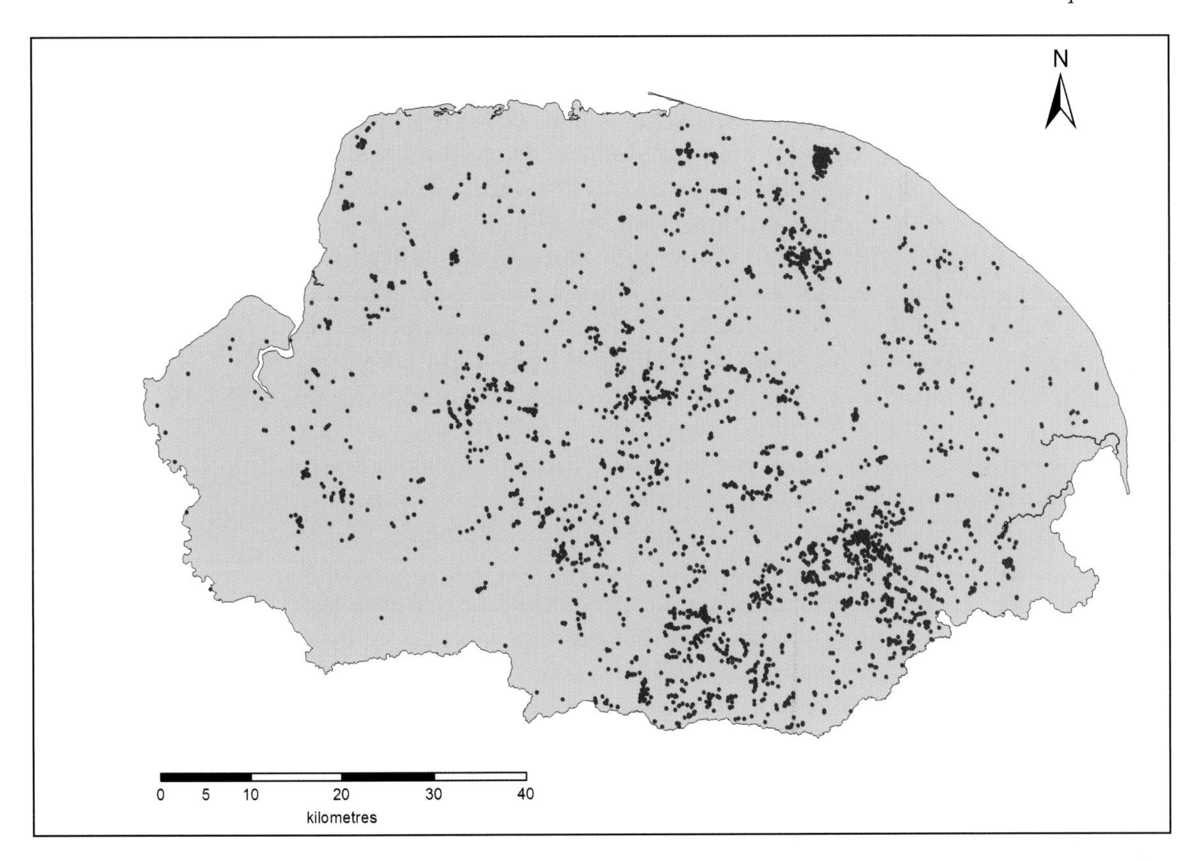

FIGURE 7. The distribution of ancient and traditionally managed trees in Norfolk.

rare and those in which they are common, and for the most part the patterns discussed in the pages that follow are probably representative and reliable. How the figure of *c.*5500 recorded trees compares with the *total* number of veteran, ancient and traditionally managed trees existing in the county today is less easy to ascertain, but systematic ground checks in and around areas targeted by volunteer recorders (as well as the investigation of 'blank' areas) have failed to produce very large numbers of additional specimens. Many such trees in the county undoubtedly await discovery, but while the total figure is certainly more than four times the recorded number, it is perhaps unlikely to be much more than six times, and almost certainly less than eight. In other words, the total number of old, ancient and traditionally managed trees in Norfolk, as defined above, is in all probability fewer than 40,000.

The location of old trees can be looked at in two ways: in terms of the kinds of landscape features with which individual examples are associated; and spatially, in terms of their overall distribution within the county, and how this relates to aspects of both the natural environment and land-use history. The majority (nearly three-quarters) of recorded specimens are hedgerow trees, with smaller numbers growing free-standing in pastures and meadows. In addition, while few very old trees are to be found – for reasons to which we shall return – within the county's ancient woods, a number of former wood-pastures, often lying hidden within areas of later planting, have now been discovered in the county, mainly as a result of this programme of research. A significant number of old trees are also to be found in 'designed landscapes' – that is, in parks, gardens and burial grounds. Here they were valued for their individual symbolism or appearance, or for the way they enhanced a view or contributed in some other way to a landscape design. They might be grown as individual specimens on lawns or in parkland, or grouped into clumps, belts, avenues and other kinds of formal planting. To some extent there is thus a distinction between the trees found in farmland and old wood-pastures – which were mainly functional, serving as a source of wood, timber, fodder and perhaps fruit – and those which were primarily ornamental in character. But such a simple dichotomy, like all simple dichotomies, falsifies. Many trees were planted with both beauty and utility in mind, especially in the eighteenth and nineteenth centuries, when large landowners embarked on ambitious schemes of afforestation. The famous belt of trees planted in the 1760s at West Tofts in south-west Norfolk, which encompassed both the park and the home farm, was on the market along with the rest of the estate in 1780, and the sales catalogue praised the beauty of the trees and the way they enhanced the prospect. But it also commented on their monetary value: the belt contained no fewer than 600,000 trees, 'which, in the course of a few years, will at least be worth a Shilling a Tree, and consequently amount to Thirty Thousand pounds ...' (NRO MC 77/1/521/7). The planting of trees in hedges could likewise result from mixed motives, at least when carried out by large landed estates in the course of the eighteenth and nineteenth centuries. As James Brown put it in 1861:

> The planting of hedgerow trees is generally done with a threefold view – namely, that of raising useful timber in the country, without occupying much breadth of land exclusively for that purpose; the producing a degree of shelter for stock and crops in the adjoining fields; and the giving the country a clothed and ornamented appearance. (Brown 1861, 395)

In addition, many trees which began life as primarily functional features, such as pollards in hedges, were later incorporated within ornamental landscapes and managed thenceforth largely or solely as objects of beauty (Muir 2005, 186–90; Williamson 1998, 135–6). Indeed, in most Norfolk parks the oldest trees originated as hedgerow pollards growing in the working landscape which the parks replaced in the course of the seventeenth, eighteenth and nineteenth centuries. Lastly, in all periods, and at all social levels, people have valued trees growing beside roads, on commons and in hedges, especially the older and larger specimens, for their beauty and associations, sometimes maintaining them well past economic maturity. Most old trees thus refuse to fall simply into 'functional' and 'ornamental' categories, an observation which is of crucial importance when we come to consider where, and why, our oldest trees have been allowed to survive to the present.

In terms of their geographical distribution, ancient and traditionally managed

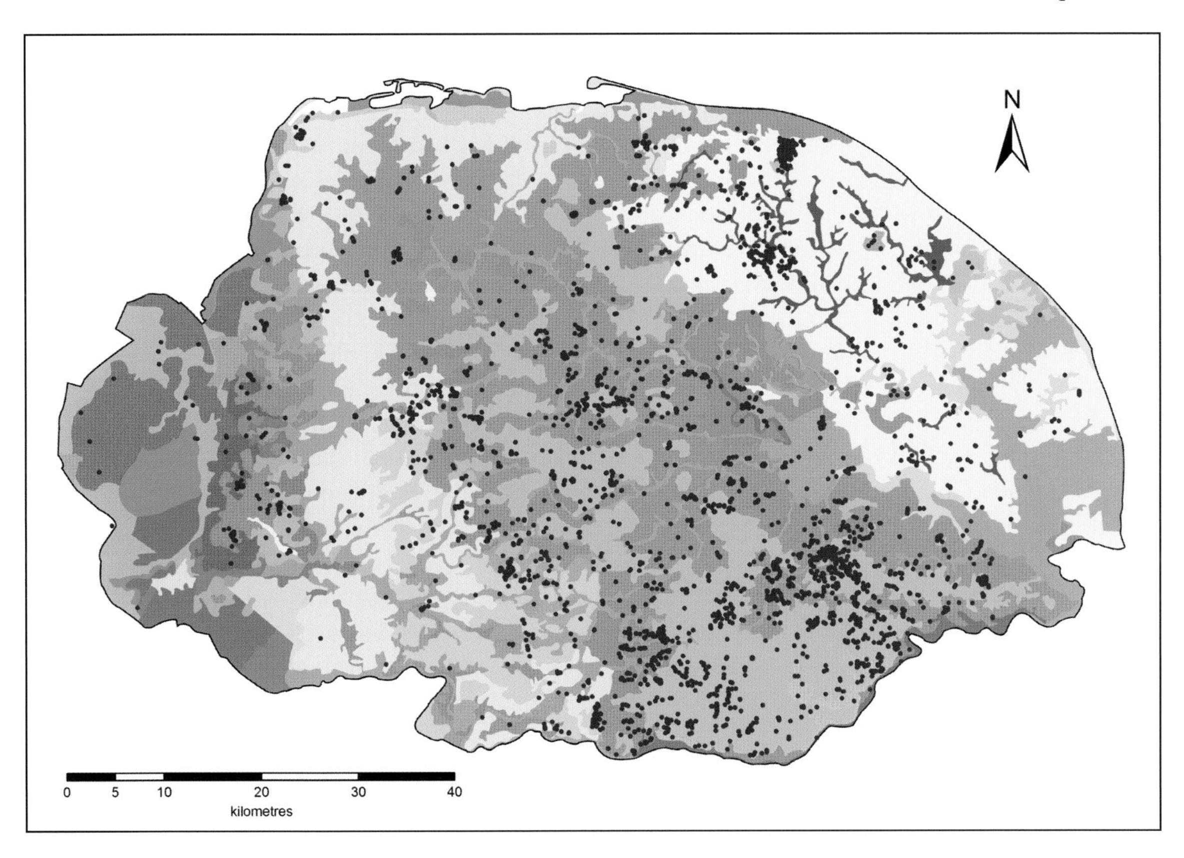

FIGURE 8. The distribution of ancient and traditionally managed trees, and principal soil types, in Norfolk.
KEY: see Figure 2.

trees are noticeably concentrated in the centre and south-east of the county (Figures 7 and 8). This part of Norfolk is characterised by poorly draining soils formed in chalky boulder clay, especially those of the Beccles and Burlingham Associations (Hodge *et al.* 1984, 117–22, 132–7). In contrast, such trees are less common on the light fertile loams (the Wick 2 and 3 Associations) in the north-east of the county, and almost unknown on the chalk soils of the west (the Newmarket 1 and 2 Associations (Hodge *et al.* 1984, 265–9)). But while soil character may have an effect on the longevity of certain species (see below, pp. 74–6), it is unlikely that this pattern mainly reflects environmental factors. This is because the next most common soil types on which old trees can be found, after the poorly draining clays of the Beccles and Burlingham Associations, are those of the Newport 4, Worlington and Barrow Associations, which are freely draining and formed in acid sands (Hodge *et al.* 1984, 107–11, 277–9). There are other interesting anomalies. Ancient trees are thus a fairly common feature of river flood plains and damp ground beside streams, but they are rare in the more extensive wetland areas of the Fens (in the west of the county) and the Broadland marshes (in the east). Moreover, while on the whole early trees are common on the Norfolk claylands, over significant tracts of them, especially in the south, none are recorded. In other words, while in *general* terms the distribution of ancient trees is clearly related to soil type, the relationship does not in the main appear to be a direct one, related to the intrinsic character of the soils themselves. Instead, as already intimated, the relationship was largely (although not entirely) indirect. Different kinds of soil were exploited by human societies, over the centuries, in different ways, some of which allowed for the early establishment of trees (in wood-pastures or hedgerows), and their survival there to the present, while others did not. Yet while the *overall* distribution of old trees is clearly determined by economic, agrarian and social factors, the distribution of particular species does seem to be more related, in a direct manner, to the environment. Old beech trees, for example, are seldom found on the heavier clays; old hornbeams are seldom found anywhere else.

Species of ancient tree

There is currently much debate about the character of the natural landscape of England before the start of large-scale clearances for farming and settlement around *c.*3500 BC. In particular, ecologists and others disagree over how dense the original tree cover may have been. An older generation believed that before the Neolithic period England was almost entirely covered with dense woodland, except for the highest mountains, above *c.*750m. Even areas which we often think of as having a 'natural' vegetation of which trees do not form a significant part, such as heaths and many areas of moorland, were once wooded. During the 1950s and 1960s there was some modification to this view as it became apparent, through the examination of pollen preserved in lakes, deep ponds and peat deposits, that some alteration had been made to the woodland cover even before the arrival of the first farmers. Hunter-gatherers in Mesolithic England, like those living in other places and times, burnt areas of woodland, presumably to facilitate hunting, and in some areas this may have opened up quite wide tracts of ground (Dimbleby 1962). Nevertheless, the established view was otherwise maintained: the natural vegetation comprised closed canopy woodland which was only gradually cleared through the Neolithic, Bronze and Iron Ages (Rackham 1976, 40). Many archaeologists and historical ecologists still believe this account, but it has recently been challenged in some respects. The Dutch ecologist Franz Vera has suggested that the grazing of large herbivores, such as the bison and auroch, prevented the development in the post-glacial period of dense and continuous forest, so that much of the English landscape in the pre-Neolithic period resembled open, park-like country, with relatively limited areas of denser trees (Vera 2002a and b). This argument, moreover, has important implications for how we manage areas for nature conservation today: 'In places where large ungulates including large grazers … could roam, the primaeval vegetation was not a closed canopy forest. It was a park-like landscape with a very high diversity of biotopes and therefore a very high diversity of wildlife' (Vera 2002b, 201).

A number of ecologists have even advocated the 're-wilding' of areas of the landscape, and the re-creation of these natural wood-pastures, in order to maximise biological diversity. But there are problems with Vera's views, which cannot be dealt with in any detail here. In a recent review of the arguments and evidence, Kathy Hodder and her colleagues commented: 'We agree that the openness of the [pre-Neolithic] landscape and the role of large herbivores have both been underplayed in past discussions, but conclude that Vera's argument – that the bulk of the lowland landscape was half-open and driven by large herbivores – is not currently supported by the evidence' (Hodder *et al.* 2009, 12). One particular problem is that we do not know the density of large herbivores in the remote past and it is quite probable that their numbers were kept in check by human predation (which had, after all, at the end of the last glaciation been responsible, in whole or part, for the complete extinction of a number of species, including the mammoth and woolly rhinocerous).

Whatever the precise character of the vegetation of the county before large-scale clearance in the Neolithic – whether tree cover was continuous or interrupted by large areas of open ground – we know a fair amount about its composition. Out of nineteen pollen cores published from Norfolk showing the pattern of vegetation just before the arrival of farming, small-leafed lime (*Tilia cordata*) was the most common tree in nearly two-thirds of cases and the second most common in a further two. Lime was principally accompanied by hazel and oak on the lighter soils and by hazel, ash and elm on the heavier ground (Rackham 1986b, 161). Other species, including pine and yew, were also present but in smaller quantities.

What is striking about this range and balance of species is that it bears only a tangential relationship to that exhibited by the old or traditionally managed trees found in the Norfolk landscape today – the subject of this book. Small-leafed lime is now a very rare tree, largely restricted to a small number of ancient woods (especially in north-central Norfolk, such as Hockering Wood). It is virtually unknown as a farmland tree, although it occurs as a shrub in a small number of hedges. It is true that lime may, for a variety of reasons, be over-represented to some extent in the pollen record, but the contrast is remarkable nevertheless. Oak (principally *Quercus robur*) is by far the most common tree recorded as ancient or traditionally managed in the county, amounting to around 73 per cent of the trees mapped on Figure 7. While it was certainly a significant component of the wildwood, especially on lighter land, oak's overwhelming dominance is clearly the consequence of human factors. Ash is a more common species than oak in Norfolk today, if we include not only fully grown trees but also the countless examples growing as shrubs or young specimens in hedges, or as coppice stools in ancient woods. It was also a prominent component of the natural vegetation across most of East Anglia. Nevertheless, for reasons to which we shall return, it is rare as a veteran. It constitutes only around 5 per cent of the trees recorded in the county as large, old, or traditionally managed.

There are roughly the same number of large, old beech trees as there are of ash in the county, while sweet chestnut (*Castanea sativa*) is the fourth most common of the recorded 'veterans', accounting for just under 4 per cent of examples. Sweet chestnut, which is not indigenous, was probably introduced to Britain in the Roman period (Rackham 1986a, 54–5). In some districts it has become naturalised, and common in the countryside. This is not the case in Norfolk where it does not occur (for example) as an element of the coppiced understorey in ancient woods. Nor is it found, except very rarely, as a tree of hedgerow and farmland. It is largely a tree of gardens and parks, where it has evidently been planted in some numbers since at least the seventeenth century. The status of beech in this respect is uncertain. It does not seem to have ever been a common tree in the East Anglian countryside, does not occur in ancient coppice-with-standards woods and is only rarely found as a hedgerow tree. With the exception of a remarkable concentration in north Norfolk, which appears to be the remains of extensive wood-pastures in the area around Felbrigg, virtually all the recorded examples are again from parks and gardens. It is noteworthy that Hannah Moor, who travelled extensively in Norfolk in the 1780s, remarked that the examples growing in the park at Houghton were the first she had seen in the county (Ketton-Cremer 1957). As we shall see, there are doubts about whether beech is, in fact, indigenous in this part of England.

These four trees – oak, ash, beech and sweet chestnut – together make up around 87 per cent of the known ancient or traditionally managed trees in Norfolk. A further four species together account for another *c.*6 per cent of such trees: hornbeam, yew, black poplar and lime (mainly *Tilia X Europaea*, the common lime, rather than *Tilia cordata*, the native variety, which, as already noted, is present in only tiny quantities in the county). Another four – horse chestnut, maple, sycamore and willow – make up a further *c.*1 per cent. The remaining *c.*5 per cent of the old and traditionally managed trees recorded in the county is made up of around thirty different species, including hawthorn, apple, alder, elm,[1] small-leaved lime, cedar of Lebanon, sequoia, London plane and holm oak.

While some of these trees are indigenous (hawthorn, apple, elder, hornbeam, maple, willow and black poplar), many are again naturalised or exotic species, introduced by man from elsewhere. Horse chestnut (*Aesculus hippocastanum*) was thus brought to England at the start of the seventeenth century; sycamore does not appear to be native in this region although it might be in the north-west of Britain (Harris 1987; Denne 1987); both, like *Tilia X Europaea*, are in Norfolk mainly trees of parks and gardens, rather than of the working countryside. Other species recorded as 'veterans' are of uncertain status. Yew is unquestionably an indigenous tree but all the old or large specimens recorded in

[1] It is likely that, had they not been destroyed by a virulent form of Dutch elm disease, which blighted East Anglia (and all other areas of England) in the 1960s and 70s, ancient examples of elm would also have been numerous. The elm has largely disappeared as a tree although it remains common as a hedgerow shrub, suckering vigorously and often forcing out other species for long lengths of hedge. Only when it begins to grow into a true tree, with fully-formed bark, does it fall victim to the elm beetles *Scolytus scolytus* and *Scolytus multistiatus*, which carry the fungus *Ceratocystis ulmi*, the cause of the disease (Brasier and Gibbs 1973).

the county are from gardens, parks, churchyards and burial grounds. Whether these are derived and descended from the native populations or were introduced (in the post-medieval period) from elsewhere after the latter had died out locally remains unclear. In a similar way, Scots pine was certainly a component of the natural vegetation of Norfolk on light soils, and especially in Breckland, but it appears to have died out in the Roman period. There are no certain references to it in medieval documents, and it appears to have been re-introduced to the county in the seventeenth century.

There is thus a rather loose relationship between the oldest trees found in Norfolk and the county's natural vegetation. Oak is far more prominent among the former than it was in the latter; small-leafed lime was common in the wildscape but hardly features at all as a veteran. A very large number of the other trees recorded as old, large or traditionally managed were either certainly, or very probably, introduced (or re-introduced) into the county at some point in the past. We often think of trees, and especially old trees, as part of the natural world, and of course in many senses they are. But our arboreal heritage mainly reflects human decisions, and thus cultural and economic factors: the presence of particular kinds of tree, in particular locations, is a consequence of choices made by individuals or communities in the past. In the working countryside trees which were particularly useful as a source of wood or timber were encouraged over those which were less so, while in parks and gardens certain species were favoured because they were individually beautiful, fashionable or exotic, or because they worked effectively as structural elements in a design when grouped together, and/or because they grew reasonably quickly. The Norfolk countryside, like that of the rest of lowland England, has been exploited, managed and cultivated for thousands of years, so we should not be surprised to find that there is no very direct link between the 'natural' vegetation which came to dominate the region in the post-glacial period and the kinds of species of tree which are common in the landscape today, and which have been common during the last few centuries. Our heritage of old trees is, in large measure, a human at least as much as a natural one. And it can thus only really be understood by first looking briefly at the wider history of the local landscape over the last thousand years or so.

Norfolk's landscapes: the Middle Ages

Soils and drainage appear to have had a decisive influence, as already noted, on the distribution of ancient and traditionally managed trees, with the bulk of examples coming from poorly draining clays or, to a lesser extent, from areas of particularly sandy, acidic soils. Soils, drainage and related factors also have a demonstrable influence on the distribution of different species of ancient trees. In part that influence has been a direct one, for while it is true that the most common tree in Norfolk, pedunculate oak, grows equally well on almost all kinds of soil, other species found in the county display clear preferences. Even if they will tolerate a wide range of soils and conditions in the early years of growth, they will attain a significant age only on a more restricted range. These preferences are sometimes obvious, as with species such as willow or alder, which are largely restricted (even as young specimens) to damp ground. Sometimes they are more subtle, as with species such as beech, hornbeam or maple, which appear to attain a significant age only in certain kinds of environment. But perhaps more important than any direct influence, soils have had an indirect – and often very indirect – effect on the distribution of old trees. This is because particular soils and environments, in all periods, have favoured the development of particular kinds of farming system, which in turn encouraged, or discouraged, the maintenance of existing trees or the establishment of new ones. For example, some kinds of farming encouraged, or even necessitated, the early enclosure of land with hedges, a factor of particular importance given that most of our farmland trees grew, and still grow, in hedges. And because farming systems were one of the factors which determined land ownership patterns, they helped to dictate the size and number of ornamental parks and extensive gardens in any area – another important repository of ancient and veteran trees in the county.

We may begin by making a broad distinction between the heavy boulder-clay soils which are mainly found in the centre and south-east of Norfolk, and the light, leached and generally less fertile soils found in Breckland (a distinctive region in south-west Norfolk), across north-west Norfolk (the 'Good Sands', to use its traditional name) and along much of the north Norfolk coast (Hodge *et al.* 1984) (see Figure 2). The claylands comprises a low plateau dissected at intervals by the valleys of the rivers Tas, Tiffey, Yare, Chet, Waveney and their principal tributaries. On the sides of the valleys the clay soils are sandier, better drained and slightly acidic, most being defined by the Soil Survey as falling within the Burlingham 1 and 3 Associations. The level plateaux between the main valleys are characterised by the poorly draining, more difficult soils of the Beccles Association. In general, these level tablelands were most continuous, and less dissected, in the south of the county, and more dissected in the south-east and towards the centre (Hodge *et al.* 1984, 117–22, 132–7).

Throughout the claylands medieval settlement was fairly dispersed in character (Williamson 2006, 153–9). As well as (in some places, instead of) villages, there were numerous isolated farms and small hamlets, mainly clustered around greens and commons which were especially large and numerous where the tracts of plateau clay were most continuous and extensive. A proportion of the farmland in these districts always consisted of enclosed fields which had been directly reclaimed from the woodlands and waste, but 'open fields' – in which the holdings of farmers were intermingled as small unhedged strips, and farming was subject to some measure of communal organisation – were also widespread, and in most areas probably dominated the landscape (Skipper 1989; Postgate 1973). Nevertheless, significant numbers of hedges seem to have existed in these districts from an early date, along roads, around the edges of the open fields and the small fields reclaimed directly from woodland, and defining the perimeters of the greens and commons.

These soils, even the heaviest of the clays, had to some extent been farmed and exploited in prehistoric and Roman times. But large areas, on the plateaux especially, had never been cleared of woodland, and many areas were abandoned for settlement in the course of the fifth and sixth centuries, after which woodland regenerated over them (Davison 1990). The ancient woods which were a prominent feature of these districts in the Middle Ages, together with wooded commons and deer parks, represented the tattered and much-altered remnants of these wooded tracts after they had been mainly cleared to make way for fields and farms in the period between the eighth and the thirteenth century. Managed private woodland, as we shall see, generally came to be located towards the margins of clay plateaux, beside the principal valleys; the larger commons, in contrast, tended to be found towards the centres of the main clay masses. It is usually assumed that these latter areas, which had once been wooded, degenerated to treeless pastures by the end of the Middle Ages under sustained grazing pressure. As we shall also see, recent research casts some doubt on this.

The light lands in the north and west of the county were rather different. Here, the geology comprises a low chalk escarpment dissected by rivers and streams. The higher ground is capped not by clay but by acid sands. Medieval settlement was less scattered than on the claylands, usually clustering in rather straggling villages. Almost all of the cultivated land, moreover, consisted of extensive open fields, generally organised on a more rigorously communal basis than those of the clays, and each usually containing the intermixed holdings of a larger number of farmers (Postgate 1973; Bailey 1989). There were relatively few hedges. Beyond the fields there were often tracts of heathland and occasionally areas of chalk grassland. Both were particularly extensive in Breckland, in the south-west of the county and extending into north-west Suffolk, a district which is distinguished by particularly dry and often acidic soils which are formed in sandy Aeolian drift overlying chalk or, in places, boulder clay. This is the most agriculturally marginal part of East Anglia, idiosyncracies of climate compounding the problems posed to cultivators by acid and infertile soils: frosts have been recorded in every month of the year (Hodge *et al.* 1984, 27–34).

Most of these light soils were sandy and acidic; some were more calcareous; but all were easily leached of nutrients, which had to be systematically replenished. This was achieved by grazing flocks of sheep on the heaths or harvest residues by day, and closely folding them on the open fields by night, when they lay fallow, before the spring sowing or when the crop was very young, so that they dunged or 'tathed' the land. Because the light soils were fairly easy to cultivate they had been attractive to early farmers and, according to conventional wisdom, their woodland had largely disappeared by the Middle Ages. Some was cleared to make way for fields, the rest had degenerated – under intense grazing pressure – to open heath. Once again, as we shall see, recent discoveries cast some doubt on the latter contention, suggesting that areas of wood-pasture survived in places into medieval and even post-medieval times. And while there were few hedgerow trees in the area, stray references in medieval documents make it clear that timber and pollards existed not only on commons but also in and around village closes and beside high-status residences. Nevertheless, areas of enclosed woodland, managed as coppice-with-standards, were much rarer in these districts by the end of the Middle Ages than they were in the south-east of the county.

Breckland and north-west Norfolk were not the only areas in which tracts of heathland existed. Isolated heaths could be found scattered through the claylands in the centre of the county, associated with localised lenses of glacial gravel. Of greater significance was a more continuous band of heathland, associated with glacial gravels as much as sands, which extended intermittently from Norwich to the north coast, interspersed with areas of cultivated land. Here some areas of ancient woodland (such as Haveringland Wood) could be found, and the presence of medieval deer parks further hint that a greater amount of woodland survived here into historic times than was the case in Breckland, although by the time the earliest maps were made, in the seventeenth century, the heaths were largely treeless. To the east of this arid district, in east Norfolk, lies an extensive tract of well-drained, fertile or moderately fertile soils dissected by broad river valleys – those of the Wensum, Bure, Ant, Thurne and their tributaries. This was light land, but more fertile that the chalks and sands in the west of the county. It was the most densely settled area of medieval Norfolk, its landscape – like that of the claylands – dominated by extensive if irregular open fields, but with a settlement pattern slightly less scattered in character, with fewer and smaller tracts of common land and with relatively little enclosed, coppiced woodland (Campbell 1981). The low-lying valleys contained areas of peat which were occupied by rough grazing, poor-quality meadows and reed and sedge beds, and by the end of the Middle Ages by extensive lakes – 'broads' – formed by the large-scale extraction of peat for fuel. The lower reaches of these valleys joined and merged in a wide tract of marshland, often loosely referred to as the Halvergate Marshes, lying to the west of Yarmouth. To the south, this area was flanked by the landscape of the southern claylands already described.

Another, much more extensive, wetland area existed in the far west of the county. Although normally referred to simply as 'The Fens' or 'Fenland', this level tract, which extends far into the neighbouring counties of Cambridgeshire, Lincolnshire and Suffolk, in fact comprises two quite distinct landscapes. Strictly speaking, Fenland is the name for the peat levels which comprise the southern portion of this district. Except for a few limited schemes of drainage and reclamation around their margins, these remained an extensive area of damp common, grazed and cut for peat, reeds and rough hay, until the seventeenth century (Darby 1983; Harris 1953). The silts and clays lying to the north, beside the Wash, were, in contrast, colonised from early medieval times and contained large villages with great parish churches (Silvester 1999). This area, which was traditionally known as Marshland, was by the thirteenth century densely settled and cultivated in extensive arable open fields, with long strips separated by drainage ditches.

Lastly, we should note the complex band of mixed geological formations, older than and underlying the chalk, which outcrop in a narrow band in the west of the county, separating the Fens from Breckland and, extending northwards, beyond King's Lynn, form a distinct band of countryside lying below the chalk escarpment and beside the Wash. Here the soils were very varied, but in general were dominated by poorly draining clays in the south, in the area between Downham Market and King's Lynn, and by acid sands and gravels to the north of the latter town. The entire district was interspersed with many broad valleys containing wide tracts of low-lying wetland, forming in effect a continuation of the Fens to the west. As elsewhere in the west of Norfolk, settlement was largely clustered in poorly nucleated, somewhat sprawling villages and the farmland largely lay in extensive open fields, accompanied by commons of various kinds.

The progress of enclosure

Few of the trees surviving in the county today date back to the Middle Ages. Most were established only as the varied landscapes we have just described underwent significant changes in the period after *c.*1400. In the course of the fifteenth and sixteenth centuries, in Norfolk as elsewhere in England, farms generally became larger and more specialised in character. In areas of light soils the existing 'sheep-corn' farming systems persisted, and for a long time the land remained open and largely devoid of hedges, and therefore of hedgerow trees. But where soils were heavier and more fertile, on the clays in the south and east of the county, the story was different. Here, farmers often came to specialise in dairying and bullock fattening and much of the land was gradually laid to pasture (Holderness 1985; Skipper 1989). The open fields were slowly removed by the process of 'piecemeal' enclosure, in which landowners gradually bought, sold and exchanged strips through a series of private agreements, amalgamating and then hedging them. The open-field systems found in these areas were particularly susceptible to this kind of informal, gradual enclosure. This is because each portion of the arable usually contained the strips of only a few proprietors, often neighbours dwelling in a small hamlet, so that few private deals were required to consolidate holdings. Nobody objected much to the change anyway, because agriculture had never really operated on highly communal lines. Piecemeal enclosure, because it involved the gradual establishment of hedges along the margins of contiguous groups of strips, tended to preserve in simplified form the older boundary patterns of the open fields, in which the strips were usually slightly sinuous in plan, many taking the form of a shallow 'reversed S' caused by the way that the ploughman moved to the left as he approached the end of

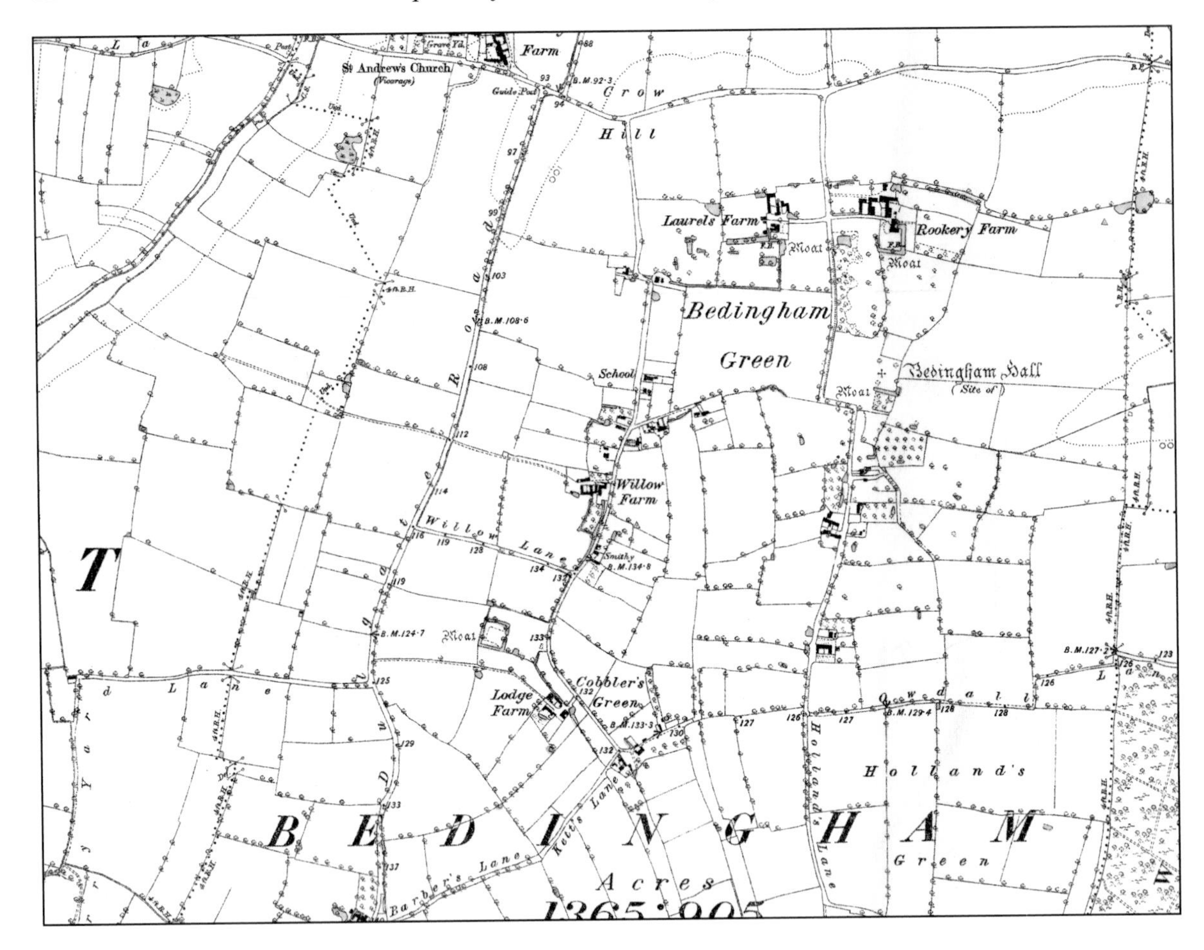

FIGURE 9. The landscape of early enclosure in south Norfolk, with scattered settlement and small fields with irregular or serpentine boundaries, as depicted on the late nineteenth-century 6-inch Ordnance Survey map.

the strip, in order to avoid too tight a turning circle (Eyre 1955). These distinctive shapes were fossilised by piecemeal enclosure, and still dominate the landscape across much of the south of the county (Figure 9). The claylands thus carry, for the most part, an 'ancient countryside', with irregularly shaped fields and winding lanes, and with hedges which are partly of medieval but mainly of early post-medieval date. Sixteenth- and seventeenth-century writers referred to this as 'woodland' countryside, not so much because it was well endowed with woods but rather because it was already full of hedgerow trees (Williamson 2006, 153–65). Nevertheless, while open fields in these areas steadily contracted through the post-medieval centuries, large areas of common land survived into the early nineteenth century because it was difficult to enclose these through informal, piecemeal methods.

While open fields had largely been removed from the claylands by the start of the eighteenth century, in the areas of lighter land to the east and west the situation was more complex. Some parishes, generally those owned by large landowners, were enclosed in the course of the sixteenth and seventeenth centuries, but it was only in the period after 1700 that the open fields began to disappear on a large scale, together with the great tracts of heath and sheepwalk which accompanied them (Wade Martins and Williamson 1999, 34–43). This development was associated in part with the adoption of the new farming methods of the 'agricultural revolution', which involved the widespread cultivation of crops such as turnips and clover and the adoption of new breeds of sheep. It was difficult to introduce modern farming practices within the old medieval framework of communal rotations, while the introduction of the new crops rendered the grazing provided by the heaths largely redundant. Open fields were thus removed wholesale, and much of the heathland and other grazing was put to the plough. While piecemeal enclosure was important in removing open arable in some places, especially in the east of the county, for the most part the complex and extensive field systems, as well as the great common sheepwalks, found in these light soil districts could be removed only by more formal and legalistic methods, which involved all the landowners in a parish agreeing to remove all the open fields and commons at a stroke, redistributing holdings in the form of new private fields laid out by surveyors and bounded by ruler-straight hedges (Wade Martins and Williamson 1999, 34–46). Many enclosures of this type were brought about by parliamentary acts in the later eighteenth and early nineteenth centuries, acts which were also used to remove residual areas of greens and commons on the claylands, where, as noted, the open arable had already normally been extinguished long before (Turner 2005). While pockets of rectilinear fields defined by species-poor hawthorn hedges can thus be found throughout the county, such boundary patterns dominate the light lands, especially in the west and north of the region, but also to some extent in the east and north-east. Indeed, in the latter districts the open fields often survived well into the nineteenth century, largely because a plethora of small owners could not easily come to an agreement to enclose and – such was the money to be made from growing grain on this rich land, largely free from fold courses and communal restrictions on farming – had little incentive to do so (Bacon 2000). The old contrast between the landscapes of the light, 'champion' lands and the more enclosed and bosky 'woodland' countryside of the claylands thus remains deeply etched in the modern countryside, although now as the contrast between the planned-looking, rectilinear countrysides created in the period after 1700 by lawyers and surveyors (Figure 10), and the ancient landscape of irregular fields, narrow lanes and species-rich hedges.

The late eighteenth and nineteenth centuries also saw dramatic changes in the landscape and land use of other districts. In the 1640s and 1650s the Dutch engineer Cornelius Vermuyden, employed by a consortium of investors headed by the Duke of Bedford, had created new watercourses and undertaken other large-scale engineering work intended to improve the drainage of the southern peat Fens, or the Bedford Level, as it came to be known (Darby 1983; Taylor 1973, 188–203; Taylor 1999, 146–9; Harris 1953). But this period did not see the creation of the kind of intensive arable landscape which we see in this area today, with ploughland spreading uninterrupted to the horizon. That came later.

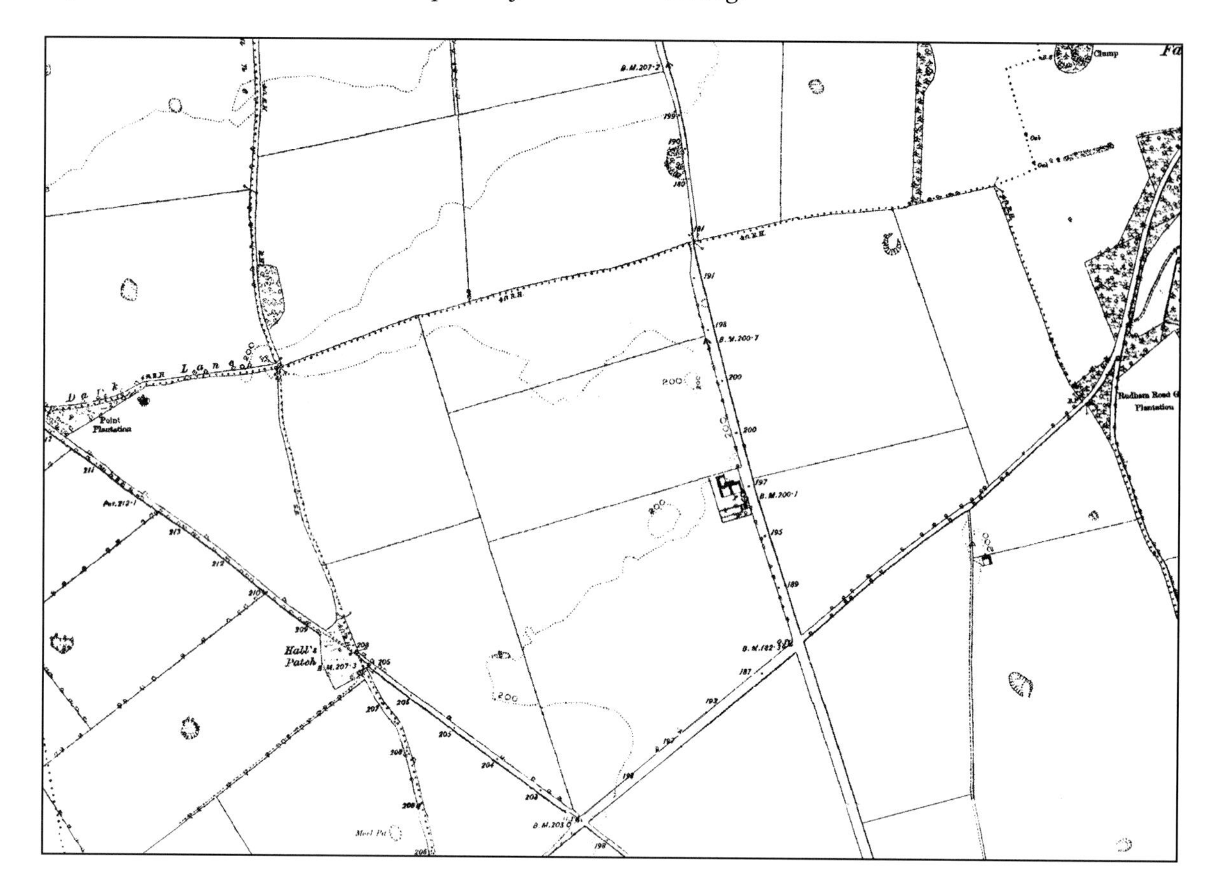

FIGURE 10. The landscape of eighteenth- and nineteenth-century enclosure: large, rectilinear fields near Sculthorpe in north-west Norfolk (as depicted on the late nineteenth-century 6-inch Ordnance Survey map).

In the seventeenth and eighteenth centuries the improvers faced continuing problems, for as the peat became drier it tended to waste and contract, falling below the level of the surrounding watercourses. The water in the drainage ditches which bounded the individual parcels of land would no longer flow out into the rivers and major artificial channels. The construction of innumerable drainage windmills remedied this problem to some extent but it was only in the nineteenth century that the use of steam pumps allowed large-scale arable land use, a development which also affected the long-settled northern fens, or Marshland. By the 1840s around two-thirds of this formerly pastoral area was in tilth (Wade Martins and Williamson 1999, 53).

We have greatly simplified a complicated story, but the essential feature of Norfolk's landscape history, crucial to any understanding of the character of its trees, is this. In the Middle Ages the claylands in the centre and south of the county contained more woodland, and already more hedges, than the lighter lands of the west and east. But the extent of this contrast increased steadily from the fifteenth to the early eighteenth century, as piecemeal enclosure of open fields proceeded, leading to a steady proliferation in the numbers of hedges and hedgerow trees. Piecemeal enclosure also affected the north-east, and to some extent the west, of the county, but the latter in particular remained a more open landscape until the eighteenth or nineteenth centuries. Looked at in this way, it is easy to understand, in general terms, the distribution of old and traditionally managed trees in the county (see Figure 7). Older trees are most common in the old-enclosed districts of the claylands simply because the bulk of the county's old trees are found in hedgerows. The north-east of the county, an area of light but fertile land which experienced also a measure of early enclosure, contains a thin scatter of such trees. But on the light lands in the west of the county, and especially on the chalky soils where open-field arable (as opposed to heathland) dominated the landscape until the eighteenth century, old trees are comparatively rare.

But there was another factor, which served to complicate this simple dichotomy between the light lands and the heavy: patterns of tenure and ownership. In the Middle Ages most large manors and castles had elaborate gardens attached to them, and wooded hunting parks, usually in isolated locations, were a common feature of the landscape. But it was only in the post-medieval centuries that ornamental parks developed as a setting for great mansions, and that the landscape in the immediate vicinity of lordly residences began to be designed on a large scale for aesthetic reasons. Such designs featured numerous new trees which were planted – as we shall see – in a variety of styles which changed over the centuries. Parks and pleasure grounds also incorporated pre-existing trees from the working landscapes which they replaced. By the end of the eighteenth century extensive designed landscapes could be found throughout the county, but in general they were less common on the heavier plateau clays, on the more fertile soils of the north-east and in the Fens, all districts in which small proprietors rather than large landowners occupied much of the land. Large estates with parks and mansions could certainly be found in these areas, but generally in smaller numbers (Williamson 2006). The prominence of small proprietors in these districts had a rather different but equally important influence on our legacy of trees, for such individuals continued, in many cases, to practise pollarding on their holdings long after the practice had been abandoned by large landowners and their tenants.

This rather extended discussion of the wider landscape history of the county of Norfolk may appear, to some readers, to be a digression away from the main subject of this volume. It is not. The distribution of old and traditionally managed trees, and of trees of different kinds, is strongly correlated with soils and topography. But that relationship is both to a large extent indirect, operating via systems of farming and patterns of land ownership, and extremely complex.

The recent history of farmland trees

So far we have discussed the history of the Norfolk landscape in the post-medieval centuries simply in terms of the processes which encouraged the proliferation of trees, such as enclosure and the establishment of extensive parks and gardens. But the current distribution of old trees in Norfolk, as elsewhere in England, is also the consequence of patterns of structured removal. Trees have always been lost from, as well as added to, the landscape, but this probably occurred on an increasing scale from the later eighteenth century. The 'agricultural revolution' of the eighteenth and nineteenth centuries did not only involve a transformation of the landscape on the light lands of Norfolk and Suffolk through enclosure of open fields and the reclamation of heaths; it was part of a wider process of agricultural change which ensured that, during the course of the later eighteenth and nineteenth centuries, East Anglia became what it is today: an intensively arable region, the bread-basket of England (Wade Martins and Williamson 1999, 205–9). This process involved, in particular, the expansion of cereal farming in the Fenland and on the clays in the south of the county, both districts which had been characterised, in the seventeenth century, by livestock farming and were largely laid to pasture.

The conversion of the latter region to large-scale arable husbandry not only involved the widespread adoption of improved systems of land drainage and the enclosure – usually by parliamentary act – of most of the surviving areas of common land but also ensured that in many places changes were made to the *existing* pattern and character of boundaries, especially where land was owned by larger proprietors, or formed part of extensive landed estates. The densely treed, irregular, 'bosky' landscape of the claylands came under sustained attack from improvers keen to maximise cereal yields. The small and irregularly shaped fields were inconvenient for ploughing, while tall hedges and their trees shaded out the crop, robbed the soil of nutrients and took up large areas of potentially productive land. Hedges were removed and fields amalgamated on a substantial scale, old hedges were realigned and replanted, and the density of hedgerow trees was often drastically thinned. Pollards were regarded with particular hostility (Wade Martins and Williamson 1999, 61–9). Their dense heads spread a deep pool of shade, and they were looked on by agricultural 'improvers' as relics of backward peasant farming. Pollards perhaps also offended 'polite' taste because they were as 'unnatural' as the topiaried trees in formal gardens, which were now out of fashion among the rich and educated (Thomas 1983, 220–21). Either way, it was a prejudice which seems to have continued throughout the following century, affecting to some degree all parts of the county in which established farmland trees existed in any numbers.

As early as 1791 a government enquiry into the state of the nation's timber supplies included the question: 'Whether the Growth of Oak Timber in Hedge Rows is generally encouraged, or whether the grubbing up of Hedge Rows for the enlarging of fields, and improving Arable Ground, is become common in those Counties?' (Lambert 1975, 708). The concern seems to have been fully justified. Randall Burroughes, a keen agricultural 'improver' who owned and farmed a substantial acreage at Wymondham on the claylands of south Norfolk in the 1790s, recorded his activities in a detailed journal. Every winter we find his men busy stubbing out hedges and taking down old pollards. In the last two weeks of 1794, for example, Burroughes described how 'Elmer & Meadows began to through down & level an old bank in part of the pasture between little Bones & Maids Yards'; reported that 'the men were employ'd in stubbing a tree or two for firing & other odd jobs'; described how 'some ash timber' was cut down, and how 'the frost continued very severe so much so that … the men employed in throwing down old hedgerows found the greatest difficulty in penetrating the ground with pick axes' (Wade Martins and Williamson 1996, 47–8). Thick, outgrown hedges were hacked out and replaced with others, straighter and neater in form, or removed entirely to amalgamate fields. In 1796 a small pasture, Whinns Close, was ploughed up and Burroughes reported how 'The trees … at the west side of the Whinn Close were stubb'd at the rate of three pence a tree and the ground firing. A beginning also was made in stubbing the hedge between Burtfield & Whinn Close the price 6d per rod & the firing. … The drainers also continued making their progress in Woolseys as did Bairn & his son in levelling the hedge bank between the Burtfield & Whinn Closes' (Wade Martins and Williamson 1996, 29). The grubbing of hedges and removal of old pollards continued on some scale right through the Victorian 'high farming' period and in 1887 Augustus Jessopp, vicar of Scarning in mid Norfolk, bemoaned how 'The small fields that used to be so picturesque and wasteful are gone or are going; the tall hedges, the high banks, the scrub or the bottoms where a fox or weasel might hope to find a night's lodging … all these things have vanished' (Jessopp 1887, 6).

In spite of such changes, the Norfolk countryside in the late nineteenth century could still boast far more trees than today. The Ordnance Survey first edition 6-inch (1:10560) maps, which in Norfolk were surveyed mainly in the 1880s, purport to show the location of every farmland tree with a girth greater than two feet: a stupendous undertaking which has not since been matched by the Ordnance Survey or anybody else. In reality, a number were omitted because the scale of the maps precluded the depiction of more than one specimen per 15m or so of hedge line, but the maps nevertheless provide a good overall impression of the number of trees, and of how this varied from district to district. The greatest densities were to be found on the poorly draining Beccles Association soils of the level clay plateaux in central and southern Norfolk. Here there were invariably more than 250 farmland trees per square kilometre, mostly but not exclusively in hedgerows; often over 300; and in some places more than 400 – well over double the numbers which we might find in the area today. This was in spite of the fact that these soils also carried significant amounts of ancient woodland (reducing the area occupied by farmland, and thus the average hedge length). This density reflected earlier landscape history. In spite of the eighteenth- and nineteenth-century changes just described, this was still an area of relatively small fields, mainly created by piecemeal enclosure, in which pastoral farming had been the most important activity before the end of the eighteenth century and in which much of the land was owned by small estates or owner-occupiers. Away from the very heaviest plateau clays, on the lighter clay loams of the Burlingham Association on the sides of the major valleys, the density of farmland trees was generally lower, with normally between 200 and 250 trees per square kilometre. Although cattle farming had also become important here in the early modern period, arable farming had continued to be of some significance on this lighter land and farmers had thus, perhaps, been less happy to see their hedges crowded with trees because of the effects of their shade on growing crops.

Similar densities could be found on the fertile loams in north-east Norfolk, another district characterised (for the most part) by large numbers of small proprietors. But, in general, lower densities of farmland trees were to be found on the lighter soils, where land had generally been enclosed from open fields and heaths in the later eighteenth and early nineteenth centuries, and was often owned by large landed proprietors. This was partly because there were fewer trees in the hedges here, but also because, as the fields produced by late enclosures were generally larger than those in areas anciently enclosed, there were fewer hedges anyway. The lowest densities per square kilometre were on the very poorest of these soils, on the former heaths to the north of Norwich and in Breckland. The figures from these areas are to some extent misleading because of the extent of plantations and shelter belts established by large estates in the wake of enclosure in the nineteenth centuries, which ensured that many fields were bounded on one or more of their sides by woodland rather than hedges. But they were also low on the better, more calcareous soils on the edge of Breckland, and in northern and north-west Norfolk generally, most areas having between 100 and 200 hedgerow trees per square kilometre. What is perhaps most surprising is the situation in the Fens of west Norfolk, now a largely treeless landscape. Although the Ordnance Survey 6-inch maps show that the southern peat fens, enclosed and drained in the course of the post-medieval period, were poorly treed in the later nineteenth century, this was not true of the northern silt fens, reclaimed and farmed since medieval times. Indeed, the density of farmland trees here was in many places well in excess of 250 per square kilometre. Most specimens grew beside the numerous drainage dykes which surrounded the fields, but some small areas of densely timbered pasture could also be found.

It is often suggested that, at the time that the Ordnance Survey 6-inch maps were being surveyed in the later nineteenth century, the rate of hedge and tree removal in the countryside was beginning to decrease. A serious agricultural depression began in the late 1870s and continued – with only brief interruptions, principally during the First World War – up until the outbreak of the Second World War in 1939. Falling farm incomes led to a neglect of hedges and boundaries after a century or more of particularly intensive management and sustained change. In Oliver Rackham's words,

> The period 1750–1870 was, on the whole, an age of agricultural prosperity in which hedgerow timber almost certainly decreased. The period 1870–1951 was, on the whole, an age of agricultural adversity, in which there was less money to spend on either maintaining or destroying hedges. Neglect gave innumerable saplings an opportunity to grow into trees. (Rackham 1986, 223)

In Norfolk, at least, matters appear to have been more complicated than this. It is clear from a comparison of the numbers of hedgerow trees shown on the late-nineteenth-century Ordnance Survey first edition 6-inch maps, and the numbers appearing on the vertical air photographs taken by the RAF in 1946, that decline continued steadily. On the claylands the decrease varied in sample kilometre squares from 5 per cent to as much as 50 per cent, and averages around 30 per cent. On the light lands the average decrease was about the same, but there was more variation from place to place, with some areas experiencing virtually no change in the number of trees and others seeing a decline of as much as 70 per cent. On the more fertile loams, especially in the north-east and east of the county, the density fell by anything between 25 per cent and 70 per cent (average 50 per cent). The most striking reductions, however, were in northern Fenland, which by 1946 had already gained its present, rather treeless, appearance. In one grid square, to the west of the village of Wiggenhall St Mary Magdalene, 441 trees are shown on the Ordnance Survey 6-inch map of 1884: by 1946 there were around 50. There is, it must be emphasised, a fair degree of subjectivity behind these figures – it is not always easy to count trees accurately from high-level vertical air photographs. There is little doubt, however, that in Norfolk, if not elsewhere in the country, the density of farmland trees declined, rather than increased, in the first half of the twentieth century.

There were a number of reasons for this, some specific to particular districts. In the Fens, for example, the drastic reduction was probably in large part the consequence of environmental change. To judge from the small number of survivors, the trees depicted on the 1880s Ordnance Survey maps were mostly willows, growing beside the drainage ditches which bounded the individual parcels of land in this watery landscape. They had presumably been planted, for the most part, in the seventeenth and eighteenth centuries. The improvements in drainage technology after 1800, especially the adoption of large steam pumps, allowed the water table to be lowered enough to permit the land to be cultivated as arable and by the 1880s probably three-quarters or more of the area was in tilth. But the lowering of the water table left many of these trees with their root systems well above the level of the waterlogged soil, and they gradually perished. More generally, even before the start of the Second World War some farmers, especially small owner-occupiers in the Fens and the north-east of the county, had begun to use tractors. The 1941 National Farm Survey, for example, shows that in the Fenland parishes of Terrington St John and Walpole St Peter there were twenty-eight and thirty-eight tractors respectively: one farm had as many as five (TNA: PRO MAF/745/281 and 742/256). The use of machinery in turn encouraged the further amalgamation of fields, the removal of boundaries and thus the loss of hedge timber. Mosby thus described in 1938 how in north-east Norfolk 'The fields are of medium size – 20 to 30 acres, but there is a tendency in some areas to enlarge the fields by removing the intervening hedge. Where this has been done the farmers, particularly those who use a tractor plough, have reduced their labour costs' (Mosby 1938, 203–4).

Loss of farmland trees was also related to important changes in the pattern of land ownership in the county in the late nineteenth and the first half of the twentieth century. As farming incomes declined so too did the fortunes of the large landed estates which, by the end of the nineteenth century, owned extensive tracts of the county, even on the southern clays. Estate owners relied for much of their income on farm rents and, with agriculture in the doldrums, these tended to fall. The rental income from the Blickling estate, for example, declined from £11,685 in 1877 to £9,893 in 1892: a major recalculation of rents in 1894 resulted in a further reduction of over a third to £6,018 (NRO MC3). On the Kimberley estates the half-yearly rents went down from £5,909 in 1880 to £3,144 in 1901 (NRO KIM 7/8). Large landowners were also adversely affected by Death Duties, introduced in 1894 and raised to 15 per cent by Lloyd George and subsequently, in 1919, to 40 per cent on estates valued at more than £200,000 (Thompson 1963, 325, 330). By the 1920s many estates were being sold and broken up and their constituent farms purchased, in many cases, by their tenants. Contemporaries often commented on the decline in the number of trees in the countryside, associating it directly with the need for large landowners to realise capital to pay Death Duties or to service their accumulating debts. Lilias Rider Haggard described in the 1930s how 'the wholesale cutting of timber all over the country is a sad sight, but often the owner's last desperate bid to enable him to cling to the family acres …', noting elsewhere: 'The other day, going past a well-known and well-loved place, I was hit like a blow in the face by a scene of complete desolation – every tree gone' (Haggard and Williamson 1943, 97).

The purchase of estate farms by former tenants was likewise frequently accompanied by wholesale felling, carried out to recoup some of the purchase price. Even where estates remained intact tree numbers often declined as hard-pressed landowners capitalised on their assets. Moreover, hedgerow timber was the landowner's property, not the tenant's, who generally resented the effects that they could have on crops in the adjacent fields. 'Farmers are, generally speaking, averse to the growing of hedgerow trees … and declare that they are ruinous to their crops, both from the great shade they produce, and from their harbouring vermin in their vicinity' (Brown 1861, 396). In these difficult times, when there were often problems in finding tenants for farms, landlords were more sensitive to such complaints than they had formerly been. Lilias Rider Haggard, who owned a small estate in south Norfolk, described:

> A consultation about the always difficult question of tree cutting on the farm. This particularly affects the arable fields, where the farming tenant has cause for some complaint. Decided somewhat sadly that some dozen small oaks must come out before the sap rises, or next autumn when the crops are off. (Haggard 1946, 73)

There is, moreover, little evidence that this was a period in which numerous young hedgerow trees had the opportunity to grow unheeded to maturity in the way that some authorities have suggested. Most trees in Norfolk hedges are oaks, which do not easily set seed and flourish in a hedge. And such evidence

FIGURE 11. Marjoram's Farm, South Walsham, in *c.*1900. It is often suggested that the great agricultural depression of the later nineteenth and early twentieth centuries led to the neglect of hedges and a consequent increase in the numbers of hedgerow trees. Most contemporary photographs, however, suggest that for the most part the Norfolk countryside continued to be intensively farmed and field boundaries well-maintained.

as there is, including contemporary photographs, indicates that most Norfolk hedges continued to be maintained with enthusiasm during the depression years, in part because this was a relatively cheap way of ensuring reasonable crop yields (Figure 11). Tall, outgrown hedges both shaded crops and provided a home for rabbits, an increasing pest during the depression years.

Whatever the true extent of hedge and tree removal in the Norfolk countryside in the years leading up to 1939, the war years unquestionably witnessed an acceleration, as the intensity of cultivation was increased to meet food shortages, as extensive areas were was given over to airfields and as timber was felled on a large scale to meet the needs of the war effort. Nor did the cessation of hostilities end this onslaught. Peace was accompanied by an extended period of food shortages and rationing which lasted well into the 1950s. First the national government, and latterly the European Economic Community, introduced a range of subsidies to increase levels of production in order to feed a hungry population: the long agricultural depression was over. Moreover, increased levels of state support were associated with other important changes in the organisation of farming. The widespread adoption of tractors and combine harvesters in the post-War years and the low cost of artificial fertilisers ensured that arable farmers no longer had the incentive to keep animals, even horses, on their holdings. The numbers of livestock fell drastically, and on many farms hedges were no longer required as stock-proof barriers. Moreover, as large machines work most economically in large fields, hedges were increasingly seen as a nuisance rather than a necessary part of the agricultural environment. They began to be removed wholesale (Baird and Tarrant 1970). Bulldozers and mechanical diggers made this a relatively easy task; so too did government subsidies for hedge removal, which began to be paid in 1957.

FIGURE 12. Pollarded oaks at Beetley. The hedge in which these magnificent trees originally grew was grubbed out in the 1960s, but the accompanying ditch, and a strip of grass, remained. When hedges were removed the adjacent fields were usually amalgamated and timber felled. The hedge has now been re-planted.

All this took a terrible toll on farmland trees, and especially on the older trees. Although some were often allowed to remain in the middle of fields after hedges had been removed, these usually failed to thrive, the open soil accelerating evaporation and starving them of water, their roots damaged by the repeated passing of the plough. Older trees were especially vulnerable in this way. Only where boundaries remained as strips of grass after hedges had been removed did old trees have a reasonable chance of survival (Figure 12). Moreover, the rates of hedgerow loss were, in general, highest in the anciently enclosed landscapes on the claylands in the south-central parts of the county, where the majority of ancient trees were to be found, especially where the land was most level and most suitable for conversion to prairies. The wholesale removal by farmers of hedges and the trees they contained in this area explains many of the marked *lacunae* in the distribution of old trees which we have already noted on the southern claylands.

It was not only the changes in the farming landscape in the course of the twentieth century which led to a sustained reduction in the numbers of farmland trees, and especially of older trees; there were also a significant decline in the numbers, and condition, of great gardens and parks. In addition to the difficult financial straits in which the owners of landed estates found themselves after the long agricultural depression leading up to the Second World War, their traditional role as leaders of the local community was challenged by a whole raft of social and legislative changes, especially the Local Government Acts of 1888 and 1894, which ensured that county affairs were no longer in the hands of county magistrates and justices of the peace who were largely drawn from the ranks of the landed elite, nor parish affairs under the control of Parish Vestries dominated by the local squire. As radical opposition to established landed wealth increased through the century, culminating in the Labour Party victory of 1945, there was little to be gained by advertising status through large mansions, which were increasingly expensive to maintain, or elaborate gardens and grounds. The great age of the country house was over and demolitions increased steadily, culminating in the 1950s, as many owners despaired of restoring homes badly damaged by wartime military occupation. Although most country houses survived their grounds were often ploughed, in whole or part, and timber removed accordingly.

All in all, the twentieth century was not a good time for old trees. But pressure abated from the late 1970s. The rate of hedge destruction seems to have declined, although in part because on some farms there were few hedges left to remove. Mounting opposition from conservationists and the general public, and the resultant withdrawal of government subsidies for hedge removal in the late 1970s, also played their part. By 1982 government policy had begun to change further, and under the Farm Capital Grant Scheme subsidies were made available not only for planting new hedges but also for renovating outgrown, neglected ones. Subsequent agro-environmental schemes, including Countryside Stewardship, also encouraged landowners to look after their heritage of trees and hedges. Moreover, new owners and established ones alike began to value and restore their parks and gardens, whether or not mansions had been found new uses as, for instance, wedding venues or country house hotels. Their efforts were encouraged by a variety of government initiatives, including Heritage Tax Exemption schemes and, once again, Countryside Stewardship.

Although vast numbers of old trees were thus lost from the landscape in the course of the twentieth century, especially in the south of the county, substantial quantities, as we have seen, still remain. And their significance and meaning – what they are doing in particular places, and why they have survived there – can only be understood against the complex history of the Norfolk landscape, briefly outlined in the previous pages.

CHAPTER TWO

Dating Trees

Approaches and methods

A crucial step in understanding the changing character, use and meaning of trees in the landscape would be to find a reliable way of dating them – not least because local opinion often attributes impossibly early dates to large or prominent examples. Oaks scarcely more than two centuries old are often said to have been planted by Queen Elizabeth, or to celebrate the defeat of the Armada – or, more vaguely, to have been established 'by the monks'. But dating individual trees is no easy matter. Even if we could fell, or core, every specimen we were interested in and count the annual growth rings this would not help us much in dating the oldest examples, for most trees more than three centuries old (and some younger than this) have lost their earliest rings through decay. Many, indeed, are almost completely hollow. The ideal would be to find a reliable method of estimating, if only broadly, the age of a tree from its size, and more particularly from the diameter or circumference of its trunk. Any number of factors can affect a tree's height, or the extent of its canopy. But, in the words of the arboriculturalist Alan Mitchell, 'the circumference of the bole of any tree must increase in some measure during every year of its life. The age of a tree is thus some function of the circumference alone' (Mitchell 1974, 25).

A number of researchers have, accordingly, tried to formulate a way of relating the circumference of the trunk (or bole) of a tree to its age. Mitchell himself suggested that, as a rough rule of thumb, the age of a free-standing tree growing in good conditions is related to its circumference measured at waist height in such a way that, on average, each inch of growth represents a year of age. In metric terms, a tree with a girth of a metre should be around forty years old. For a tree growing in woodland, competing with neighbours for light and growing tall rather than wide, the relationship would be two years for each inch of growth; for a tree in an 'intermediate' location, such as an avenue or a small clump, the figure would be a year and a half (Mitchell 1974, 25). Mitchell's rule is a useful one, which can give a very approximate indication of age for many indigenous trees, but, as he himself acknowledged, it fails to take full account of variations in the rate of growth exhibited by trees of different species and trees growing in different soils and other environmental circumstances. Moreover, as he was also well aware, trees do not grow at a uniform rate throughout their lives; rather, they put on girth faster in youth, more slowly in middle age and then much more slowly in senescence. This latter observation – that growth slows markedly in later life – is by no means new. Indeed, as early as 1841 the Norwich nurseryman James Grigor – whose book *The Eastern Arboretum* discussed and detailed the most beautiful, biggest and oldest trees which the author knew of in Norfolk – criticised estimates made of age by the French scientist Louis Bosc: 'The calculation of M. Bosc, perhaps, would approximate nearer to the truth, if he allowed a greater diameter to the oak in the earlier stages of its growth, and, on the contrary, a much less diameter during the period of old age' (Grigor 1841, 338).

John White's more complicated dating method makes a more systematic allowance for all these factors (White 1997 and 1998). In particular, White explains in some detail how, and why, growth rates change over the life of a tree. In the early phase of growth the amount of food available to the tree is expanding as its canopy develops. But because the diameter of the trunk is increasing rapidly the annual growth rings tend to be fairly even in width, and wider than those produced in maturity. When the crown of the tree is fully formed – usually after 40 to 100 years, depending on species – the amount of food produced by the foliage remains relatively stable, unless the tree is damaged in some way – by extreme weather conditions, for example, or insect infestation. The amount of new wood added to the tree each year will, therefore, likewise remain fairly constant but, because the size of the tree is still increasing, this volume will be more thinly spread, over a larger area. As a result, the annual growth rings will tend to gradually narrow. In old age, as the canopy dies back and the tree sustains increasing physical damage, less food is available and the rings narrow even further; in the oldest trees they may be no more than 0.5mm (amounting to twenty rings per centimetre) in width. Very big, old trees can thus put on extra girth each year by such a small amount that they appear to be virtually 'flat-lining'. As White warns, however, not all trees behave in precisely the same way.

Some species groups, such as oak and chestnut, keep faithfully to the three phases of growth format outlined above. Other trees, however, do not. 'Pioneers such as poplar, willow and alder frequently have a productive but short formative period and then go straight into senescence. Birch, which is relatively short-lived, tends to have an extremely brief middle period. Yew, on the other hand, lives a charmed existence. It can return to formative rates of growth at almost any stage in its very long life. It may be stimulated by a boost of plant food from branch layering, or by vigorous regeneration after catastrophic damage' (White 1998, 2).

White, having examined a large number of felled trees, was able to produce tables showing patterns of growth in different species – the age at which, for example, specimens of particular species will shift from 'youth' to 'middle age'. This is important because, as already noted, in old trees the original 'core' growth has usually been destroyed. The tree is hollow, and the point of this transition cannot therefore be directly ascertained even if it is felled. The rate and duration of 'core' growth, however, is greatly affected by local factors, such as whether the tree has been growing in an open situation or hemmed in by neighbours, and White emphasised the importance of examining the landscape context of the tree. His method thus takes particular note of such things as whether a tree growing in a park or a hedgerow is truly free-standing or has its canopy touching that of a neighbour.

After a long period of study White was thus able to produce a dating method which involves a number of distinct stages. Firstly, the species of the tree needs to be noted and its context and growth pattern assessed. The method assumes (based on many years of observation of felled stumps and annual measurements) that the 'core size and the speed of early growth within a given species group on a particular site type' will be broadly the same. Although White provided a table indicating the likely point at which this first stage of growth would come to an end, he also emphasised the importance, where possible, of compiling local site tables based on observations of felled trees. The crucial difficulty in this first stage of the calculation lies in deciding the conditions in which the tree first grew.

> This is critical because all the subsequent calculations of age depend on the core age and ring width indicated. Observed conditions at the site of the tree must be thorough but treated with caution. These probably did not prevail many years ago when the tree was young. Much will have changed since then. (White 1998, 3)

White divided the situations in which trees grew into seven general categories: 'champion tree potential' – that is, ideal growing conditions; 'good sites', where trees were sheltered but not hemmed in by other trees; 'average sites', in parkland or gardens; sites in churchyards; sites on poor or exposed ground; sites on woodland boundaries or open woodland; and sites within woodland.

Having categorised the tree's situation, the next step is to measure the diameter at breast height (DBH) – conventionally defined as 1.3m above ground level. Let us assume that the tree is an oak, and its diameter is 1.5m (and radius therefore 0.75m). If it is growing on an 'average' site the tree will pass (according to White's tables) from core growth to middle age at around 80 years. Prior to this the annual rings will have averaged 4mm in width. In other words, 0.64m of the 1.50m diametre will be accounted for by the first phase of rapid growth or, to put it another way, the area of this central core of wood, imagined as a cross-section, is around 3217 square centimetres. The total base area (the area of the total cross-section) will be 17,662.5 square centimetres, and if we subtract from this the area occupied by the 'core' of initial growth (3217), this leaves 14,445.5 square centimetres. This figure then needs to be divided by the average area of the rings in the second, mature phase of growth, which White's table tells us is, in this particular case, 79.7 square centimetres. This gives the age of the second phase of growth as 181 years, to which we have to add the 80 years of the first phase, giving a total of 261 years.

White's method, based on many years experience, is careful and meticulous, and seems to work well with comparatively young trees. The problem lies with trees older than around 300 years, which begin to sustain damage to the crown, heralding the onset of senescence and reduced growth rates and annual ring size. White's method assumes that this can be estimated by inspecting the condition of the crown, assessing how many years have passed since decline set in, and then adding this number to the calculated age (having subtracted the number of centimetres of diameter this would account for, on the rough basis of 20 years per centimetre). But with very ancient trees such assessments of past growth history are neither easy nor objective, even for the trained dendrologist. Moreover, as White is at pains to emphasise, the conditions in which a tree grew in the earlier stages of life have a crucial impact on its size. But the older the tree, the more the character of its immediate surroundings will have changed and the harder it will be to reconstruct them. As he himself noted, 'determination of site history is often a matter of some speculation'. What is now a lone example may once have been surrounded by competitors; even the character of the soils in the immediate area may have been altered, by improvements in drainage for example. There are further problems, especially concerning the effects of regular pollarding on the way in which trees put on girth. While some arboriculturalists assume that this had little or no impact on the growth of the tree, some landscape historians have suggested otherwise, arguing that it may significantly reduce the rate at which they increase their circumference, so that big pollards are actually rather older than they appear, or than we might estimate using White's method (Muir 2000).

One way of ascertaining the relationship between the size of trees and their age would be to discover sufficient examples which can be dated using historical documents. Unfortunately, we know the planting dates of relatively few individual trees, and these mainly in gardens and parks. In the wider countryside it was rare for people in the past to describe their tree-planting activities on paper, and even when they did so it is generally impossible to identify on the ground today the particular specimens to which they were referring. Although, as we have noted, the Ordnance Survey first edition 6-inch maps of the late nineteenth century show the position of all well-established trees in parks and hedgerows, only a handful of earlier maps, mostly produced by private estates in the eighteenth and early nineteenth centuries, depict individual trees, and again it is not always clear how these relate to specimens growing in the same locations today. Yet, while maps generally fail to show individual trees, they often show the field boundaries in which so many are located and, where successive maps of the same area exist, the absence of a boundary from one map and its presence on another allows an estimate of the hedge's age – broad or narrow, depending on the period of time separating the making of the maps in question – to be made. In most cases, a tree will be no older than the hedge in which it grows, so that dating the hedge provides an earliest possible date for a tree growing in it. Unfortunately, trees were often *added* to hedges long after the latter were first planted, so this approach can only take us so far. In any case, the period of time between the production of successive maps is often two centuries or more, greatly reducing the accuracy with which we can date the boundary; relatively few parishes in Norfolk have a good series of pre-nineteenth-century maps; and very few detailed maps exist at all for the period before *c.*1650.

Another line of approach relates to trees growing in eighteenth- and nineteenth-century parks, many of which were incorporated from the earlier working countryside. While hedges were removed to produce the open prospects of turf required by fashion some of the hedgerow trees were normally retained. These can usually be distinguished from those established when, or after, the park was created, either because they still grow in noticeable lines, because they are associated with the earthwork traces of the old field boundaries, or because they were pollarded – pollarding was only rarely carried out in ornamental parks, as by this stage it was considered aesthetically unappealing by the wealthy. Unfortunately, we do not know how old any such trees were when they were incorporated within a park, and while it is easy to assume that the smallest examples may have been little more than saplings, we cannot be sure of this (only a small proportion of such former hedgerow trees have usually survived within particular parks up to the present).

In short, it is difficult indeed to date any individual tree, especially those growing in farmland, from maps or documents. This said, some more general aspects of their location and distribution can suggest probable date ranges for trees of particular size and species. But it is important at this point, perhaps, to make an explicit distinction between two quite different things: the probable age of a particular tree of a particular size; and the *maximum* possible size which a tree of a particular age and species might reach. These, as already noted in our discussion of White's dating method, are not the same. A tree growing in optimum conditions will be considerably larger than one growing in less advantageous circumstances.

Oak

Oaks are by far the most numerous of the ancient and traditionally managed trees found in Norfolk, both in parklands and in the wider countryside. Most are pedunculate oak (*Quercus robur*), rather than sessile (*Quercus petraea*), which is largely restricted to a limited area in the north of the county. The sheer number of these trees allows us to make some useful observations about the relationship between girth and age, at least in the case of those planted within the last three centuries or so. In understanding this relationship two issues are of particular importance: the effects on growth rates of management, especially pollarding; and the effects of variations in soils, drainage and other environmental circumstances.

We may begin by examining the trees planted in the ruler-straight hedges established when areas of common land were enclosed or field boundaries tidied up and realigned on the claylands of south Norfolk in the late eighteenth and early nineteenth century. The largest standard oaks found in these locations can reach, on rare occasions, as much as 4.5m in circumference at waist height, more commonly 4.0m. Ring counts carried out on recently felled specimens present a similar picture. Ten examples planted in the first two decades of the nineteenth century in the clayland parishes of Great Melton, Wymondham, Marlingford and Hethersett displayed girths ranging from 3.3m to 4.2m. While we can easily demonstrate that standard trees of late-eighteenth-/early-nineteenth-century date growing on moist clays can attain girths in the 4.0–4.5m range, it is harder to show the smallest girths which trees of this age might reach because it is impossible to tell whether the smaller trees found in these straight hedges were planted when the hedges were first established or represent later additions. We can, however, be reasonably certain that most pollards growing in these hedges will be contemporary with their first planting because, while some new pollards were created even after 1900, for the most part pollarding declined in the course of the nineteenth century. Pollards recorded from straight hedges on the clays are mainly between 2.5m and 3.0m in girth (fifty-four examples) and less frequently between 3.0m and 4.0m (thirty-nine); a minority fall into the 4.0–4.5m range (seventeen).

Such as it is, this evidence suggests that pollarding may have had some slight effect on growth rates although, as noted, we cannot show whether standard trees would exhibit the same numerical ranges because we can identify only the largest of the trees planted as original components in these straight hedges. More obvious are the effects of soils on the present dimensions of trees planted in these relatively straight, relatively recent hedges. On the light soils in the north and west of the county the boundaries of this kind are, in general, slightly earlier in date than those found on the clays. Parliamentary enclosure of open fields occurred on some scale in these districts two, three or more decades earlier than the enclosure of the clayland commons. Moreover, some boundaries of this kind were created by schemes of reclamation or landscape reorganisation carried out by large estates in the middle or even early decades of the eighteenth century (Wade Martins and Williamson 1999, 34–49). It is noticeable, nevertheless, that pollards found in these hedges are actually *smaller* in circumference than those on the heavier soils, ranging from 2.0m to 3.8m, although there are fewer of them.

The evidence thus suggests that oaks planted in Norfolk around two centuries ago can exhibit considerable variations in growth, measured as circumference at waist height, from around 2.5m to as much as 4.5m. These differences result in part from environmental circumstances, in part perhaps from methods of management in the past, and in part from largely unpredictable circumstances of early growth. There is no reason to believe that any variations in dimensions established during the first two centuries of growth become less marked as the trees get older (Lennon 2009, 177–8). It is therefore interesting to examine what are, in general, a slightly older sample of trees – those growing within landscape parks created in the second half of the eighteenth century. Some of these are standards, newly established when the park was laid out; some are hedgerow trees, usually pollards, incorporated from the earlier agricultural landscape. The latter will have been of very varying age at the time but the smallest examples must presumably have been around two decades old, allowing for the two pollarding cycles necessary to create a truly 'pollarded' appearance. Such trees thus provide us with some indication of the *smallest* amount of girth trees of this kind, planted in the second half of the eighteenth century, can attain. The size of such trees can be compared with those of the largest standards, unassociated with the earthworks of former field boundaries and apparently planted when the park was first established, thus allowing us to see the *largest* size which oaks planted in the second half of the eighteenth century can reach.

Raveningham Park, on the claylands of south Norfolk, was apparently established in the early 1780s, and the smallest pre-park pollards here are between 3.6m and 3.8m in circumference. The largest free-standing parkland trees, which must by definition be slightly younger, have girths of as much as 5.5m. At Great Melton Park, created in the 1790s, the smallest pollards are 3.8–4.0m, the largest free-standing oaks 5.3m; at Gawdy, laid out in the 1780s, the figures are 4.2m and 5.0m (Williamson 1998, 234–6, 263–4, 268–9). On lighter land the trees in both categories are slightly smaller. Thus at Honing, in north-east Norfolk, created before 1791, the figures are 3.2m and 5.1m; at Earlham, just to the west of Norwich, probably laid out in the 1760s or 70s, they are 3.6m and 5.0m; at Little Dunham, perhaps from the 1770s, 2.6m and 5.0m (Williamson 1998, 247–8, 228–9). These figures suggest that oaks planted on moist clays in the second half of the eighteenth century can, on rare occasions, achieve girths in excess of 5.0m; that in general trees growing on lighter land have rather smaller girths than those on heavier clay; and more tentatively, that pollards may put on girth more slowly than standards even after they have ceased to be cropped.

These same patterns can be discerned in the relatively few oaks which can be dated, with some confidence, to the later seventeenth and early eighteenth century. The oldest trees in the great south avenue at Houghton, planted in the 1730s (which survive among much more recent replanting), thus have girths averaging 3.9m, but ranging from 3.2m to 5.0m. Here the small average size reflects the closely planted nature of the trees in question, which are separated by only 7.0m. Nevertheless, the smallest of the former hedgerow pollards which were incorporated into the park when it was expanded at around the same time are generally smaller, with girths of 3.4m, 3.5m and 3.7m. Even the smallest of the old field boundary trees incorporated into the park when it was first created, probably in the later seventeenth century (Williamson 1998, 248–9), have girths of only 3.9m, 4.4m and 4.6m. Houghton lies on light, freely draining land, as does Quidenham, in the south of the county, where a felled oak with a girth of 4.2m had a ring count suggesting it had been planted shortly after 1700. Trees of this kind of date growing on clay soils can, not surprisingly, attain slightly larger girths. Three felled oaks from heavy clays in the Wymondham area which were planted, to judge from ring counts, around 1750 had girths of 4.5–4.7m; while a pollarded oak in Morley, growing in a hedge established (to judge from the map evidence) between 1624 and 1816, has a girth of 5.2m (NRO PD3/108; C/Sce 2/5/13). A hedgerow pollard at Gressenhall, growing within an area shown as unenclosed open arable on a map of 1624, has a girth of 5.6m, but the hedge in question might have been planted at any time over the following 170 years.

Although circumstances of environment, management and early growth thus generate considerable variation in growth patterns among oaks, we can be reasonably sure of the *maximum* possible size which can be attained by trees of this species which are of eighteenth- and nineteenth-century date: for there are no unequivocal examples in the county of oaks planted after 1700 which have girths of 6m or more. There are, in all, only 170 trees of this size recorded in the county, a figure which further research would certainly double, possibly treble, but perhaps not quadruple. Of these, 118 have girths in the range 6.0–6.99m, while 52 have girths of 7m or more.[1] But it is important to emphasise that, while we may be reasonably confident that trees with girths greater than 6m are of seventeenth-century or earlier date, it will be apparent from what has been said that a great many examples with girths considerably less than this will have been planted well before *c.*1700.

This said, it is useful to examine the places where these particularly large old oaks are to be found. The first thing to note is that whereas smaller veteran and traditionally managed trees are noticeably concentrated on the claylands in the centre and south-east of the county, these massive trees tend to be more evenly spread; and that the very largest, those of 7m or above, display no obvious preference for soil type (Figure 13). This immediately suggests that their distribution is not, as appears to be the case with the generality of old and traditionally managed trees, structured by the chronology of post-medieval enclosure of open fields and commons. No examples are associated with the straight hedges created by eighteenth- and nineteenth-century enclosure and re-organisation, and few with the kind of sinuous boundary created by piecemeal enclosure of open arable in the course of the fifteenth, sixteenth and seventeenth centuries. Instead, they tend to be found in places which lay outside the main arable areas of medieval Norfolk, beyond the margins of the former open fields. A number are thus found within villages, as at Carbrooke, or within sections of villages deserted in the later Middle Ages, as at Houghton or North Elmham; in places where fragments of ancient wood-pastures have been preserved, usually within later plantations, as at Bayfield, Thursford or Stow Bedon; on heaths or former heaths, as at Wretham (Figure 14); or on flood plains, which were invariably used as meadow or pasture in the Middle Ages and later. Where they are found in existing or former hedges these tend to be of particular types. Almost all are along the sides of roads: as we explained earlier, in many parts of the county, even those in which much of the arable

[1] The remainder, for a variety of reasons, were recorded approximately only.

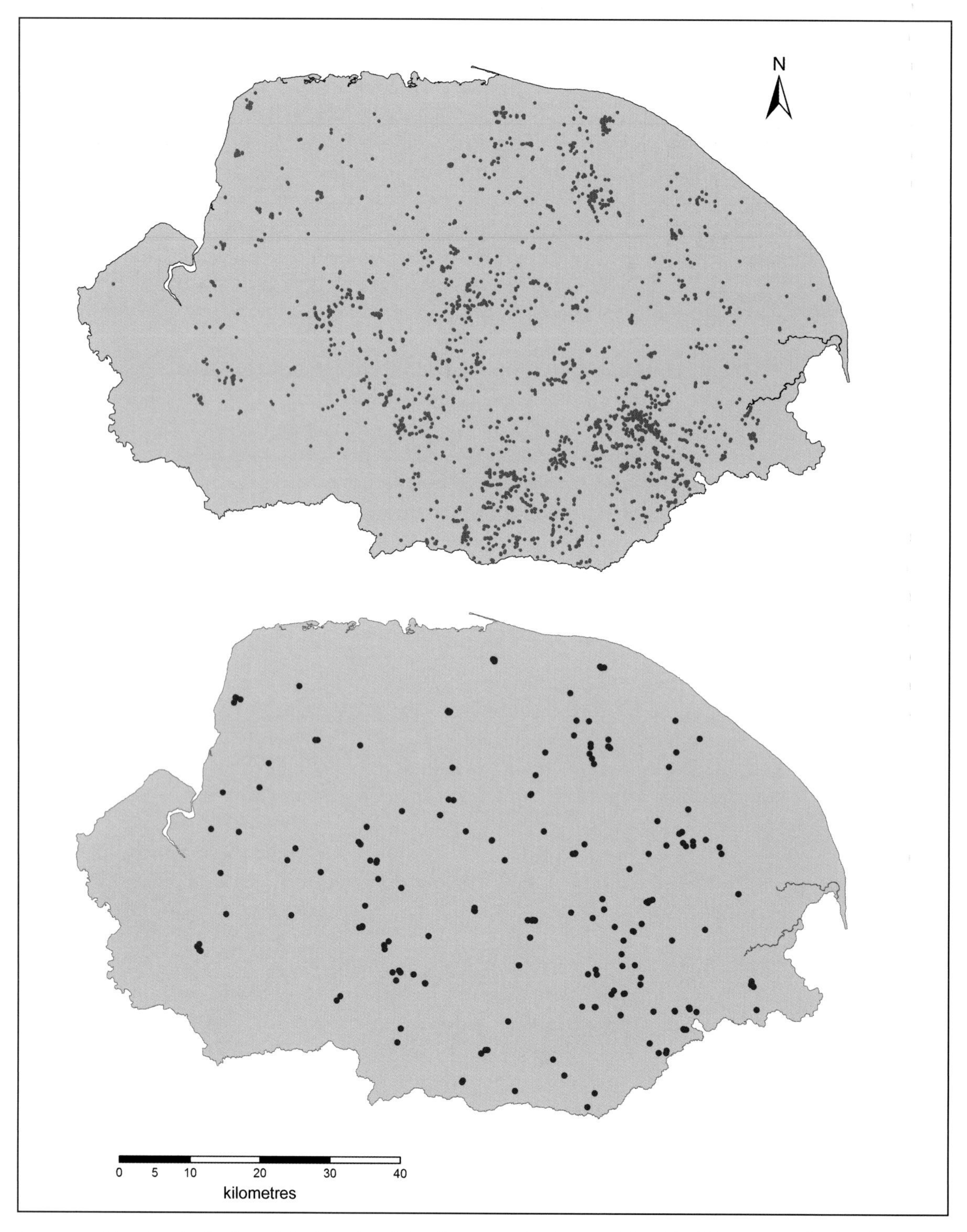

FIGURE 13. Top: the distribution of all old and traditionally managed oaks recorded in Norfolk. Bottom: the distribution of oaks with a girth at waist height of 6m or more. Note how the latter are more evenly distributed across the county than the former.

FIGURE 14. The great ruined oak at East Wretham, with a girth of more than 9m.

land lay unenclosed in the Middle Ages, roadsides were often hedged. A few are in long boundaries which once divided two open fields, as apparently at Little Dunham, or in hedges which formerly separated common land from areas of arable. Others are associated with fields which have probably always been held in severalty: either because they were reclaimed directly from the waste in the Middle Ages, such as those preserved in earthwork form within Kimberley Park; or because they were part of the enclosed ground around manor houses, as is probably the case with the ancient oaks in the park at Raveningham, all of which grew in or around fields which, to judge from the evidence of an estate map, had been enclosed by the 1620s. This latter grouping forms part of a wider category, of trees closely associated with (within *c.*50m of) the known or probable locations of medieval manorial sites, including moated sites. In all, more than two-thirds of the very large trees in the county, with girths greater than 6m, are found in one of these locations, suggesting strongly that a significant proportion are of medieval date: some at least of the remainder *may* have been similarly located, but the evidence for their original location is less strong (as, for example, where trees now grow in the middle of eighteenth-century parks and any archaeological evidence for their original landscape context has been lost). What is particularly striking, however, is that while around 60 per cent of trees with girths of 6.00–6.99m are found in one of these potentially 'medieval' locations, the figure for those of 7m or more rises to over 75 per cent, strongly suggesting that the overwhelming majority of these largest of oaks are indeed of fifteenth-century or earlier date.

One other feature of the distribution of these large oaks (with girths in excess of 6m) should be noted. A noticeably high proportion, around 28 per cent, grow on flood plains or near (within 75m of) rivers and watercourses. To this we can add a number of places where trees grow within the same distance of significant ponds, bringing the total to around 38 per cent. In addition, several major concentrations of such trees, most notably the huge specimens in the park at Kimberley, are associated with areas of particularly damp, poorly draining soil. One explanation for this association is that in such well-watered locations oaks were less likely to be weakened by periodic drought and thus more likely to survive long into senescence. But the possibility remains that a few of these trees may not be as old as they look, and that growing in these conditions has allowed them to thrive. Some of the largest trees, with girths greater than 7m, may, in fact, have been planted in the sixteenth or, just conceivably, the seventeenth century, especially those which still boast full, healthy crowns. In a similar way, a very small proportion of the specimens in the 6.00–6.99m range may conceivably have been planted after 1700.

Yet if some really large trees might be younger than they appear, others could be of considerable antiquity, for, as already noted, the growth rate of oaks, as of other trees, slows considerably in senescence. Unfortunately, there are very few trees for which we have measurements made in the past which can be confidently identified with specimens surviving in the landscape today, and even where such comparisons can be made there is always uncertainty over whether different observers measured the tree in slightly different places, or at slightly different heights. Nevertheless, the few examples we do have are instructive. James Grigor thus describes an oak growing near the home farm beside the park at Beeston St Andrew which must be the tree that still stands 'next to the pond by the farm steading'. Its girth today at waist height is 6.8m; Grigor measured it as nineteen feet (5.8m); suggesting that oaks of around 6m, even growing in optimum conditions, will add no more than around a metre to their girth in the course of the following two centuries. Kett's Oak in Ryston Park, a great ruin of a tree but still apparently healthy enough, was measured by Grigor some time before 1841, when it had a girth of just under 9.0m at two feet (*c.*0.6m) from the ground (Grigor 1841, 348), probably equivalent to around 8.5m at waist height. Today, 170 years later, its girth is around 9.0m. Some oaks, it is true, appear to sustain a greater rate of growth into old age. The ancient Winfarthing Oak (below, pp. 132–4), which finally died at the end of the nineteenth century, was measured in 1744 by the pioneer arboriculturalist Robert Marsham, and again in 1820. In 1744 the circumference was thirty-eight feet seven inches (11.8m); by 1820 it had grown to forty feet (12.2m). Growth must have been slowing markedly by this time, however, for by 1873, when it was again measured, there had been no detectable change (Grigor 1841, 354; *Mirror of Literature* 1836, Nov. 19, 345; Amyot 1874, 12, 15, 18). Its average annual growth rate had presumably dropped to the 1mm proposed by White for very ancient specimens, effectively undetectable on an uneven tree of this great size. How old might such a tree have been? It originally grew within the deer park in Gissing, which was probably established – by enclosing a portion of the surviving 'wildwood' – in the thirteenth century. It is hard to believe that the oak was not already mature at this time.

In this context, it is interesting to note the size of the largest Norfolk oaks which James Grigor, writing 170 years ago, recorded as being of significant size. One at Thorpe next Norwich, described as 'The largest oak in the village, and one which we may class with the most noted in our county', was sixteen feet three inches (4.95m) in circumference (Grigor 1841, 53). At Taverham he noted how the oaks 'seem to have found a soil peculiarly suited to there nature' and had attained 'extraordinary dimensions'. One had a trunk sixteen feet six inches (5.0m) in circumference, while another – 'which we are inclined to reckon amongst the largest in Norfolk', had a girth of nineteen feet six inches (5.9m) (Grigor 1841, 57–8). 'Magnificent' examples at Blickling had girths of nineteen feet three inches (5.8m) and eighteen feet three inches (5.6m); a 'noble' tree at Elmham a girth of nineteen feet (5.8m); while an example at Barningham Hall, which Grigor believed was 300 years old and 'now in its decrepitude', was seventeen feet (*c.*5.2m) in circumference (Grigor 1841, 101, 170, 119). A 'magnificent pollarded oak' at Seething, which he singled out for particular attention, was only 4.9m in circumference, and a nearby specimen, 'topped about a century since', was *c.*6.9m (Grigor 1841, 246). An oak growing in Ditchingham Park, which Grigor considered 'may vie with any in the county', was twenty-five feet at the base and sixteen feet at five foot, probably equivalent to around twenty feet (6.0m) at waist height (Grigor 1841, 254). Even the 'Great Oak' at Thorpe near Gunton, 'the oak of the county, one of the most extraordinary trees in this land of trees' – was no more than 6.6m in girth, while the 'Bixley Oak' – 'a tree that has been conceded to be of no common pretensions' – was a mere 5.3m (Grigor 1841, 134, 270). 'Immense pollarded oaks' in the grounds of Easton Lodge were 'generally about eighteen feet' (5.5m) in circumference, not very large compared with trees of this species found in the county today.

It is true that Grigor was perhaps more interested in trees of particular tall and luxuriant growth, rather than those with particularly large girths, although the fact that he regularly supplied information about the latter suggests that it was one of his key measures of size. It is nevertheless striking that with only three exceptions – the famous Winfarthing Oak, a massive specimen at Merton which was thirty-two feet four inches in circumference at a height of three feet, probably equivalent to around 9.5m at waist height, and one measuring 8.2m at Necton – none of the oaks discussed by Grigor had girths greater than 7.0m.

This strongly suggests that the early-nineteenth-century countryside, while it doubtless contained a great many more large old oaks than survive today, did not contain many which were larger than the biggest which we have today. In the climatic conditions and agricultural circumstances of a well-settled county such as Norfolk few oak trees have probably ever lived for more than 500 years.

The main conclusions of the foregoing discussion may be summarised as follows. The available evidence suggests that oaks planted around 1800 can, in extreme cases, attain girths of around 4.5m, while those planted around *c.*1700 can approach, but perhaps only very rarely exceed, 6.0m. Oaks with girths in excess of this are thus likely to be of seventeenth century or earlier date; but so too are some much smaller trees. Most specimens with girths in excess of *c.*7m appear to pre-date 1500, although some may be a century or more younger and, once again, some quite small oaks may be of equivalent age. In short, there is no simple and dependable relationship between age and size, and trees of the same age can exhibit very different girths, the consequence of variations in soils, past management and now irrecoverable circumstances of early growth.

Other farmland trees

There are far fewer ancient or traditionally managed examples of other species of farmland tree surviving in the county, and few of these can be even approximately dated from maps or documents. In numerical terms the second most important species, after oak, is ash (*Fraxinus excelsior*). Hardly any standard specimens of this species have attained any great size, and the majority of examples were recorded in the survey because they were pollards, rather than as particularly large examples. Only forty-four have girths greater than 4m and only nine more than 5m: the largest recorded examples were 6.0m and 6.4m. A shallow-rooting tree, and prone to rot, it is evident that ash simply does not live for very long and most examples with girths greater than *c.*5m are probably approaching the end of their lives (Figure 15). It is noteworthy that Grigor, writing in 1841, emphasised that ash trees 'of any great size' were 'rare in Norfolk' (Grigor 1841, 161). In general, ash appears to put on girth more slowly than oak. A number of examples which were planted, and then pollarded, following the enclosure of common land at Morley and Wymondham in south Norfolk at the start of the nineteenth century today have girths in the range 2.6–3.2m. Other roughly dated examples include specimens with girths of 2.7m planted in the early nineteenth century at Sprowston Lodge, just to the north of Norwich. However, it is also evident that ash, even more than oak, can display considerable and to some extent unpredictable variations in growth: of particular interest is a tree of precisely the same circumference – 2.7m – in Honeypot Wood near East Dereham, which grows on top of the blast bank raised, probably in 1941, around a wartime bomb store (Figure 16). Again, it is instructive to note the girths of specimens which Grigor, nearly 170 years ago, singled out for their great size.

FIGURE 15. An ash pollard in a hedge at Morley, on the claylands of south Norfolk,now approaching the end of its life.

One at Morton Hall was twelve feet (3.7m) in circumference; an example at Cromer was ten feet (3.1m); one at Seething, thirteen feet (just under 4.0m); and one at Kirby Cane, probably a pollard, fifteen feet (4.6m), although Grigor tells us that this specimen was measured at the base. The specimen beside the river in Kimberley Park, 'the largest tree of the species hitherto recorded in England', was around 6.7m in girth. Once again, the similarity with the present situation is striking. The implication is that, given local conditions, ash trees will only rarely exceed girths of 4m, and even more rarely exceed 5m. Trees of the latter size growing in good conditions may be around 400 years old although, as the Honeypot Wood example suggests, they might in theory be much younger.

Hornbeam (*Carpinus betula*) appears to put on girth at a slower rate than

FIGURE 16. Honeypot Wood, Wendling. This ash, with a girth of 2.7m, must post-date 1941, as it grows on top of the blast bank raised around a Second World War ammunition dump hidden within the wood.

either oak or ash. The largest example recorded in the county has a girth of 5.7m, but the second largest is only 4.8m and, in all, only thirteen have girths of more than 3m. The vast majority of the 86 trees recorded are in the 2–3m range, with twelve pollards having girths of less than 2m. These are striking trees, with their beech-like leaves and grey bark, and with trunks often flattened and twisted owing less to past management than to their unusual growth pattern – their medullary rings are large and widely spaced, so that the annual growth rings break down into discrete patches. In spite of their modest girths, these trees generally have substantial, gnarled heads, the result of many cycles of pollarding (Figure 17). The potential antiquity of specimens in the 2–3m range (the majority in the county) may be indicated by a pollarded example with a girth of 2.8m recorded from Shotesham Park, which was apparently incorporated from a hedgerow when the landscape here was first laid out in the 1780s (Williamson 1998, 277) and which can hardly, therefore, have been planted much after *c.*1750. The Norfolk hornbeams are not unusual in their small size. Some larger examples are known from elsewhere in the country, most notably Hatfield Forest in Essex, several of which have girths nearing 6m, but they are rare and probably always have been. As Loudon noted in 1840, even the largest hornbeams are 'seldom more than 6ft–9ft [*c.*2.0–3.5m] in circumference' (Loudon 1838, III, 2006).

FIGURE 17. In spite of its relatively small size, this fine hornbeam pollard in a hedge near Burston in south Norfolk is probably more than three centuries old.

Maple, similarly, seldom attains any great size in Norfolk. Of the fifty-two examples of old and traditionally managed examples recorded – again, mainly pollards with markedly gnarled heads – only nine have girths of more than 3m, the largest being a specimen of 3.6m at Wretham. There is little reliable dating evidence: an example apparently surviving from a hedgerow within Houghton Park, grubbed out in the mid-eighteenth century, has a circumference of less than 3m, although this had probably been managed as a hedgerow shrub up to this point (Figure 18). Hawthorns are in some ways similar. They seldom put on any very considerable girth, although in part this may be because the vast majority have been managed, for the majority of their lives, as hedging plants. Free-grown specimens with girths in excess of 3.6m have been noted, but examples in parks apparently left when hedges were grubbed out in the middle or later decades of the eighteenth century commonly reach only 2.0–2.5m.

Compared to these species, others found in the countryside, especially willow and black poplar (*Populus nigra* var. betulifolia), put on girth at a ferocious rate. The best evidence relates to the latter species, which is relatively uncommon in Norfolk but has received a fair amount of attention over recent years (Cooper 2006; Rogers 1993; Barnes *et al.* 2009). Although a number of recorded specimens have substantial girths – up to 7m in the case of one at East Harling in Norfolk – it is evident that the tree can grow very quickly. Specimens planted in the Earlham cemetery in Norwich, which was established only from the 1850s, already have girths of 3.7m. Two in Wymondham, growing in a field boundary laid out when the parish was enclosed in 1810 (NRO C/Sca 2/345), have girths of 3.8m and 3.9m. Another example, in the south of the same parish, has a girth of 3.8m and yet, to judge from the evidence of the Ordnance Survey 6-inch maps, must post-date 1888, at which time the area in which it grows comprised a large arable field. Specimens planted in the arboretum at Ryston Hall as late as 1904 – admittedly on particularly moist and fertile ground – have girths of no less than 3.5m and 3.6m (Ryston Hall archives). Growth is not always as rapid as this, however, especially perhaps in pollarded specimens. A pollard at Tibenham appears to have grown on the boundary of Pristow Green before this was enclosed in 1824 (C/Sca 2/300) and was thus almost certainly established before this date, yet it has a girth of only 4.0m. Nevertheless, even allowing for some reduction in growth rates in middle and later life, it is probable that even the largest trees in the county do not date from before the seventeenth century. The largest in the county, that at East Harling with a girth of 7.0m, may be the tree which was reported to John Evelyn by Sir Thomas Browne in 1663 (Rogers 1993). The dating evidence for large old willows is less good, but these, too, evidently put on girth at a comparable rate. Massive old specimens (once again some of 6.0m or more were recorded) are once again unlikely to pre-date the early eighteenth century (Figure 19).

FIGURE 18 (top). This maple in the park at Houghton in west Norfolk must be older than it looks: it is a survivor from a hedge grubbed out when the park was expanded in the 1730s.

FIGURE 19 (bottom). These magnificent willows, growing on a flood plain near Roydon, are of uncertain age, but possibly less than two centuries old.

Parkland and garden trees

We have, on the whole, rather more information with which to consider the relationship between age and size when we turn to the kinds of trees usually planted only in parks and gardens. This is because the development of these 'designed landscapes' is often (although by no means always) documented in some detail in surviving documents and maps. We have, in particular, probable planting dates for a number of sweet chestnuts, limes and beeches.

Beech trees are sometimes found in farmland, and some may represent remains of early wood-pastures, but the majority of veteran specimens grow in parks and gardens, often close to country houses, like the magnificent example – with a girth of 6.8m – beside Bayfield Hall. There are ninety-eight specimens recorded in the county with girths of 5m or more, twenty-three with girths of 6m or more and five which are larger than 7m. But even the largest of these trees are not as old as we might assume, for beech trees can evidently put on girth at a fairly rapid rate. Examples with girths of 5.7m and 5.8m growing in the park at Fring in north-west Norfolk cannot have been planted before 1807, when the hall was built on a virgin site and the park laid out around it at the expense of arable fields (Figure 20). In 1841 James Grigor noted one specimen at Wolterton, planted by Horatio Walpole around 1720 and thus only 120 years old at the time he was writing, which had already attained a girth of 4.9m (Grigor 1841, 113). Yet, like oak, beech trees can evidently put on girth at very variable rates, depending on that by-now-familiar range of factors (competition with

FIGURE 20 (below). In spite of its great size, this beech tree in the park at Fring must post-date 1807, when the house was constructed on a virgin site and the grounds laid out around it.

FIGURE 21 (opposite). A massive beech tree, with a girth of 6.8m, in an avenue at Houghton planted in *c.*1730.

neighbours, drainage, soil type and so on). Some trees significantly older than those at Fring can exhibit considerably smaller girths, such as those planted when the park at Raveningham was first created around 1780, although possibly as part of a shelter belt, which have circumferences of 4.0m, 4.2m, 4.2m, 4.8m, 5.0m and 5.2m (averaging 4.6m); or an example at Melton Constable, where a solitary survivor from the perimeter belts put in by Capability Brown in the 1760s or 1770s has a girth of 4.9m. And, once again, trees in close proximity can girth at varying rates. Three beeches, of the copper, cut-leaf and chestnut-leaf varieties, apparently planted when the grounds of Stow Bardolph Hall were redesigned by Lewis Kennedy in 1812, have girths of 5.0m, 3.6m and 3.8m respectively (Williamson 1998, 268–9, 279). The largest securely dated examples are found in the park at Houghton, in a double-planted avenue running north from the hall, which was established by the designer Charles Bridgeman in *c.*1730 (Williamson 1998, 249) (Figure 21). Only a handful of original trees still survive, and these are approaching the end of their life. They have girths of 4.8m, 5.0m, 5.8m and 6.8m (averaging *c.*5.6m). On this basis, it would appear that while beech trees planted around 1800 commonly have girths of around 4m they can easily attain well over 5m, while trees planted in the middle decades of the eighteenth century can clearly reach 6m, although they may be much smaller.

Beech not only puts on girth faster than oak but it is also much shorter-lived. It is noteworthy that many avenues, groves and clumps known to have been planted wholly or partly with beech in the early or middle decades of the eighteenth century now contain no examples of this species. Obelisk Wood in Holkham Park, for example, was partly planted with beech in the 1720s, but none now remain. In 1898 the vast size of the specimens here were a source of wonder to visitors (*Gardener's Chronicle* 1889, I, 86–7). It is probable, in fact, that in Norfolk at least few beech trees can live for longer than 300 years, at which point they may, in the right circumstances, have attained girths of 7m or more. This in turn suggests that beech is not a long-established component of the county's vegetation, for in 1841 James Grigor recorded no examples of this size, the largest being specimens with girths of fourteen feet (4.3m) at Barningham and fifteen feet (4.6m) at Kimberley (Grigor 1841, 119, 276). These are noticeably smaller than the largest examples now growing in the county, suggesting in turn that the oldest beeches that existed here in the mid nineteenth century were significantly younger than those surviving today.

Sweet chestnut is the fourth most numerous tree of 'veteran' status in Norfolk. No fewer than 180 examples with girths of over 4m have been recorded, almost all in parks, gardens and churchyards. No fewer than 23 have girths in excess of 7m, and seven have girths of 8m or more. These are vast, striking trees, but probably not of any extreme antiquity. Certainly, to judge from dated examples, trees of this species can put on girth as rapidly as beech. An avenue at Houghton planted in the 1860s or 1870s contains examples with girths ranging from 2.0m to as much as 3.8m; one with a girth of 4.6m at Sprowston Lodge near Norwich was almost certainly planted around the time the house was built in *c.*1816; while James Grigor, writing in 1841, described specimens planted on the South Lawn at Stratton Strawless, 'the nuts of which were sown in 1720', which already had an average circumference of fifteen feet (*c.*4.5m), and examples at Salhouse Hall, planted around eighty years before he was writing, which were eight feet (2.4m) in girth (Grigor 1841, 86, 305). As with beech, of particular note are the very large specimens found in some eighteenth-century parks. While some may be survivors from avenues or other early geometric planting schemes laid out around the mansion in the seventeenth century this is unlikely to be true in all cases and the examples of 5.0m and 5.4m at Heacham, for example, were almost certainly planted when, or soon after, the park there was created between 1768 and 1773 (Williamson 1998, 230–40; NRO HEA 488 256X4). Those in Weston Longville Park, of 5.5m and 6.0m respectively, cannot pre-date the construction of the new hall, on a virgin site in the 1770s, while the examples at Langley, with girths of 6.1m and 7.0m, must similarly have been planted when or after the hall was erected on what had previously been open farmland in the 1720s (Williamson 1998, 259, 284). The largest of the trees planted at Wolterton when the grounds were redesigned in the 1720s, in part by Charles Bridgeman, have girths of 6.0m, 6.1m, 7.2m, 7.9m and 8.0m.

Large sweet chestnuts are thus not necessarily as old as they appear, and even examples with girths of 8m may be no more than early eighteenth-century in date, although most specimens of 7m or more are probably seventeenth-century. An example growing in the churchyard at Hevingham, traditionally said to have been planted in 1647, thus has a girth of 7.5m (Figure 22). It had already attained 6.4m when examined by James Grigor in *c.*1841 (Grigor 1841, 91).[2] Sweet chestnuts in Great Melton Park, which were probably originally part of a formal garden of late-seventeenth-century date, had girths of between seventeen and nineteen feet (5–6m) in 1841, and now measure from 7m to 8m. The largest trees in the remains of the avenue, almost certainly of early-seventeenth-century origin, running north from Heydon Hall have girths of 7.1m and 7.8m, while the largest in the late-seventeenth-century avenue at Houghton has achieved 7.3m (Williamson 1998, 240–41, 248–9). Contemporary and adjacent specimens here are rather smaller, and in general the spacing of trees clearly has an important impact on growth. The sweet chestnuts planted in a distinctive 'fan' in the grounds of Rougham Hall in west Norfolk in the 1690s have girths ranging from 5.4m to 5.8m, but they are very closely spaced, forming in effect a small wood (Taigel and Williamson 1991, 89–90). Many specimens much smaller than this, especially those closely planted in clumps and avenues, may be of similar age, or even older.

It is noteworthy that Grigor described the example in the churchyard at Hevingham as 'excelling all we have seen in the county', although it was then,

[2] A few metres away, in the garden of the former rectory adjoining the churchyard, are a number of other sweet chestnuts, some of which may be of similar vintage, others perhaps their descendants: the largest have girths of 6.1m and 7.2m.

FIGURE 22. The great sweet chestnut in the churchyard at Hevingham, with a circumference of 7.5m, is said to have been planted in 1647.

as noted, only twenty-one feet (6.4m) in circumference. Examples at Hedenham were noted for their girths (measured at five feet above ground) of only fourteen feet (4.3m) (Grigor 1841, 254). Such figures suggest that few trees of this species were planted in the county before the seventeenth century, and that even the most massive surviving examples, like that at Hanworth Hall (now just under 8m), probably post-date 1600.

Around sixty large old specimens of common lime (*Tilia X Europaea*) have been recorded in parks and gardens across the county. In many cases dense epicormic growth around the base of the trees makes it difficult to measure their girths with any accuracy, as does the highly 'fluted' nature of some trunks. Examples in an avenue at Rougham, apparently survivors from the original planting of the 1690s, have girths of 5.4m, 5.5m, 5.5m, 5.7m and 5.8m (averaging 5.6m) (Taigel and Williamson 1991, 89–90). Four specimens growing in Kimberley Park, slightly more recent in date but open-grown for much of their lives, have girths of 5.4m, 5.5m, 5.9m and 6.3m (they were probably planted when the construction of Kimberley Hall, on a virgin site, commenced in 1712: they are certainly shown on a map of *c.*1750) (Private collection: Taigel and Williamson 1991, 71). Two fine limes growing to the north of Merton Hall, remnants of an avenue shown on a map of 1733 and which is stylistically perhaps unlikely to pre-date *c.*1650 (NRONNRO 90/2 Microfilm), have girths of 6.0m, 6.0m and 6.1m: Grigor measured one of these in the 1830s and found it had a girth of eighteen feet (5.5m) at a height of one foot above the ground, probably equivalent to a little under 5m at waist height (Grigor 1841, 350). Examples planted in the park at Raveningham towards the end of the eighteenth century have girths of 4.2m and 4.3m, but these may originally have grown in woodland; while an example in the pleasure grounds at Stow Bardolph, almost certainly planted when these were redesigned by Lewis Kennedy around 1812, has a girth of 4.2m (Williamson 1998, 279). Trees growing in a line in the grounds of Morningthorpe Manor, which were planted after 1839 (when the land they occupy was an arable field: NRO DN/TA 140) but before 1884 (when they were mature enough to be depicted on the Ordnance Survey 6-inch map), already have girths in the range 2.8–3.4m (averaging 3.2m). These figures suggest, once again, a significant degree of variability, and it is perhaps unfortunate for present purposes that most of our older examples were planted in avenues, where the close spacing has generally suppressed growth. It seems clear, however, that open-grown limes can easily attain a girth of 4m in 200 years, while the specimens in Kimberley Park suggest that they can reach 6m in 300, with trees in avenues and similar close-grown locations exhibiting considerably smaller girths. Even the oldest limes in Norfolk, like those at Merton, thus probably date to only the middle decades of the seventeenth century. Once again, it is instructive to compare these figures with those presented by Grigor in *c.*1841. A particularly large example at Barningham had a girth of twelve feet (3.7m); one of those in the Merton avenue was, as noted, just under *c.*5m; but no larger examples are mentioned, clearly indicating, perhaps not surprisingly, that the oldest limes in the county in Grigor's time were planted around the same time (the mid–late seventeenth century) as the oldest which survive today.

We thus have enough information to understand, at least broadly, the relationship between girth and age exhibited by beech, sweet chestnut and lime. We have rather less information for other species commonly planted in gardens and parks. While a few of the forty specimens of old sycamore trees recorded are from what are probably relict wood-pastures, and mature examples are sporadically encountered in hedgerows, most are found in gardens, churchyards, parks and similar locations. Few can be dated accurately, only one recorded example has a girth of over 5m, and it is most doubtful whether any Norfolk examples pre-date *c.*1750. The species can certainly put on girth rapidly. Trees flanking the drives at Benacre, just over the county boundary in Suffolk, which were unquestionably planted between 1887 and 1907 (Benacre archives) have girths in the range 2.0–2.8m in spite of the fact that they are growing very close together. Other kinds of park and garden tree may grow even faster than this. Around thirty-five horse chestnuts – a species introduced into England in the early seventeenth century, and in Norfolk widely planted in gardens and parks but seldom in the wider countryside – with girths of 4m or more are recorded from the county, the largest being specimens of 6.0m at Lexham and Watton. Again, none are probably of any great antiquity. Ring-counts of felled specimens with girths of 4.0m and 4.2m in the parks at Houghton and Kimberley revealed that they were around 150 years old, and while these might be extreme cases (examples at Sprowston Lodge, probably coeval with the building of the house in 1816, have girths of 3.5m, 3.6m and 4.5m) there is no doubt that trees of this species commonly acquire girths in excess of 3m within a century. Indeed, in some circumstances the tree can grow even faster than this: at Benacre in Suffolk thirty-three specimens unquestionably planted in the park in 1907–11 today have girths ranging from 2.0m to just under 4.0m, and averaging 2.9m. Horse chestnuts thus grow fast and die young, and it is very unlikely that many surviving Norfolk examples pre-date 1750, although those closely planted on the Ice House Mound in Houghton Park, the largest of which has a girth of 5.7m, may just do so (the mound itself was completed around 1745) (Williamson 1998, 250).

Other species of tree found in Norfolk's parks and gardens of 'veteran' status are fewer in number and, in consequence, less can be said about the probable age of specimens of different sizes. London planes (*Platanus X hispanica*) seem to grow rapidly. Magnificent examples at Stow Bardolph, planted around 1812 when the gardens were redesigned by Lewis Kennedy (Williamson 1998, 249), have girths of 3.8m, 4.0m and 4.1m; examples in the pleasure grounds at nearby Ryston Hall with girth of 5.7m and 5.9m are probably only slightly earlier in date. Even the largest examples in the county, with girths of 6m or more, are thus unlikely to pre-date *c.*1700. Many ornamental conifers, likewise, put on girth quickly. Wellingtonia (*Sequoiadendron giganteum*) is the most dramatic example. Only introduced into England in 1853 (Mitchell 1974, 86), there are nevertheless many recorded examples in Norfolk with girths of more than 4m, and some (in the parks at Catton and Langley) greater than 6m. Cedars, too, are seldom as old as their size suggests. One cedar of Lebanon (*Cedrus Libani*) in the park at Intwood, which is situated to the south of the kitchen garden and must have been planted after the latter was reduced in size and its boundary realigned in the 1880s, has a girth of well over 4m. Another example, in the grounds of Sprowston Lodge, has a circumference of 5.0m but cannot pre-date the construction of the house and the laying out of the grounds here in *c.*1816 (Williamson 1998, 278). Even an example in the grounds of Langley Hall, with a girth of 8.5m, cannot have been planted before the eighteenth century: there was no house, park or garden on the site before *c.*1720 (Williamson 1998, 259). An example planted by Robert Marsham at Stratton Strawless in 1747 had already attained a girth of twelve feet two inches (4.0m) by *c.*1840, when James Grigor measured it (Grigor 1841, 84). Today what is almost certainly the same tree has a girth of 7.1m.

Rapid growth, although not perhaps quite as rapid as the examples noted above, is clearly a feature of most other conifer species. Scots pine (*Pinus sylvestris*), unlike cedar or Wellingtonia, is native to England. It was a feature of the drier East Anglian soils in the prehistoric and Roman periods but may have died out by the Middle Ages, for there do not seem to be any medieval references to the species, and there are few certain post-medieval ones before the seventeenth century, when the plantation established in the grounds of Somerleyton Hall in east Suffolk was evidently considered something of a novelty (Freeman 1952, 523; Williamson 2000, 16–18). Few surviving examples appear to be of any great antiquity. With the exception of one huge specimen at Langley, with a girth of 5.8m, there are no recorded trees of this species in the county with girths in excess of 5m. Even the Langley example cannot pre-date 1720, when the hall was constructed on a virgin site, so most if not all large old pines are quite possibly, at most, of early-eighteenth-century date. As we shall see, the pines in the famous Breckland 'rows', which were mostly planted in the early nineteenth century, generally have girths of less than 3m, often less than 2m, owing to the ways in which they were planted and managed in early life.

In marked contrast to the species just described, apple (both crab and domestic) and pear put on girth quickly, but seldom attain a circumference of more than 2m, at which point they can be well over a century old. Again, these species will be dealt with in more detail later: suffice it to say at this point that even the most ancient specimens never exhibit girths greater than 3m.

The variability of growth rates

One point above all others stands out from the rather extended discussion and analysis presented in the previous pages. While the ages we have suggested for trees of various sizes and species are broadly in line with what might be expected from John White's dating method, our data tends to emphasise the great range of variation which can be exhibited by trees of the same date, even when growing in close proximity, as in a hedgerow or an avenue. Indeed, there are clear signs that close-planted specimens display a greater range of variability than free-standing specimens after even a short period of time. The oaks in an avenue at Morningthorpe Manor, for example, planted after 1838 (they are located within an area shown as an arable field on the tithe award map: NRO DN/TA 140) and at least semi-mature by 1884 (when they appear on the Ordnance Survey first edition 6-inch map), have girths ranging from 2.3m to as much as 3.8m. Nearby *free-standing* oaks, again planted after 1838 and mature by 1884, only have girths in the range 3.0–3.4m.

FIGURE 23. Horse chestnuts growing around the Ice House Mound at Houghton. The trees, probably planted in the eighteenth century, were originally spaced evenly and close together, almost as a hedge. They now display a great range of sizes, some of the trees having 'got away' at an early stage of their life, thus suppressing the growth of neighbours.

The success of one tree leading to the poor performance of a neighbour can often be seen. A good example can be found in the park at Houghton, where a ring of horse chestnuts was planted around the large Ice House Mound close to but not at the level of the surrounding ground. Twenty-nine of the trees survive of what was originally a closely planted continuous ring, with individual specimens spaced at *c.*3m intervals. In places, especially where this spacing still remains, the trees have rather similar girths of *c.*3.2–3.6m. But elsewhere very large trees grow next to very small ones: in an extreme case, a tree with a girth of 5.7m next to one of only 2.5m. Where one tree has 'got away', and/or has benefited from the poor growth of a neighbour, it can be more than twice the girth of that neighbour after several centuries of growth (Figure 23).

Lennon has argued persuasively that the *average* girth of trees growing in features like clumps and avenues ought to be a reasonable guide to the age of the planting as a whole (Lennon 2009, 176–7): but this would only work if we could be sure that the extant specimens constitute the majority of those once planted rather than – as is often the case – a small minority of survivors. Certainly, marked variations in the circumferences of closely spaced trees can develop quite quickly. The sweet chestnut avenue at Houghton, which now forms the drive to the east lodge but which was originally part of the complex pattern of geometric planting otherwise swept away in the 1730s, is a good example. It was extended to the north-east and partly replanted in the late nineteenth century.[3] The surviving section of original planting – ignoring what are clearly late-nineteenth-century replacements – includes trees with girths ranging from 4.2m to 7.3m and averaging 5.4m. While some of these may also be later replacements, it is striking that the trees in the nineteenth-century extension display a similar pattern of variation. The smallest tree here has a girth of 2.0m; the largest, 3.8m. Another practical problem with Lennon's 'mean girth' procedure is that parkland features such as avenues often run across an extended area of ground, and thus through and across a range of environmental conditions. The horse chestnut avenue which crosses the park to the south of Heydon Hall is a good example. A rather gappy avenue, defined by single lines of trees, is shown on this line on the 1886 Ordnance Survey 6-inch map. The surviving trees are all clearly later in date and define a more complex feature, with a single-planted southern section; a central section, in which the avenue is defined by clumps; and a northern, double-planted stretch. The southern and central sections are shown, more or less as they are today, on the Ordnance Survey map of 1907; the northern section, however, is there only shown as single lines of trees. The girths of the trees in the southern section, which grow on moist and sheltered ground, range from 3.5m to 4.6m, although the map evidence leaves little doubt that they must all have been planted in the late nineteenth century. The trees in the central clumps, planted at or around the same time, are noticeable smaller, with girths in the range 1.9–3.8m. This is more droughty and exposed ground. Those in the double-planted northern section, which must be of twentieth-century date, have girths from 2.4 to 2.9m. This section runs, for the most part, across level, moister and less exposed terrain.

Other examples of varied growth displayed by trees apparently planted at the same time in clumps, avenues and similar features have already been noted. Particularly striking are the beech trees at Houghton, almost certainly original components of an avenue of *c.*1730, which have girths ranging from 4.8m to 6.8m. There is little reason to doubt that close-planted hedgerow trees display a similar range of variability, although only in a few cases can we demonstrate that all were planted at the same time. Beech is very rarely encountered as a hedgerow tree in Norfolk so the two examples growing in close proximity in the same hedge at Bintree are presumably contemporary: they have girths of 7.5m and 5.2m respectively. There is at least a suspicion that the lines of old oak pollards encountered in many hedges may, in some cases, have all have been planted at the same time, probably when the hedge was first established, in spite of their widely varying size today, rather than some having been added to the hedge at various later dates. At Wattlefield, to the south of Wymondham,

[3] It probably dates to the 1860s or 1870s: its trees were too small to be recorded on the Ordnance Survey 25-inch survey of 1891, but are shown on that of 1906.

seventeen pollard oaks are spaced at along a roadside hedge at intervals which vary from around 3m to as much as 25m. The trees have girths which range from 3.1m to 5.2m and average just under 4m. It is possible that all were planted when the road first became hedged, perhaps in the seventeenth century, and that variations in size result from some trees having overtopped the others at an early stage of growth, with the additional complication that the death and failure of some of the trees in later life have given neighbours further opportunity to flourish.

White has emphasised the important contribution made to the final size of a tree by the conditions experienced in the early stages of growth, but for most trees growing in hedges or farmland it is hard or impossible to know what these were like, and especially how far trees may have been crowded or over-topped by neighbours. Certainly, early maps and documents suggest much variation in the density of hedgerow trees. On a property at Ickburgh in 1651 there were 6.3 per acre (NRO WLS XXVI/4 414X6); on a Beeston farm in 1761 there were around 8 trees per acre (NRO WIS 138, 166X3); at a place near Whissonsett there were at the end of the eighteenth century just over 6 trees per acre, although it was noted that a few years earlier (in 1794) there had been more, amounting to 13 per acre (NRO PD 703/45–6). At Stanfield there were 9 per acre at the end of the eighteenth century but, once again, much thinning had recently taken place and a few years earlier the figure appears to have been more like 16 per acre (NRO PD 703/45–6), a figure similar to that seen at Shipdham in the 1760s (ESRO HA 54 970/360). Much of this variation was related to the size of fields, and thus the density of hedges, on the properties in question. But the density of trees along the length of the hedges also varied greatly. At Stanfield in the 1790s there was on average around one tree for every 6m of hedge, but on a farm at Attleborough in the 1780s one for every 8m (NRO PD 703/46). More striking, however, were the variations from hedge to hedge on the same property: at Attleborough, for example, these seem to have ranged from one tree for over 13m of hedge to one for only 4m (NRO PD 703/45–6). Because only a handful of documents providing this kind of information survive, however, it is simply impossible to know how densely crowded particular veteran trees may have been in the early stages of their life. All this means that, with our older trees at least, any distinction we might make today between an 'open grown' specimen and one with close neighbours is usually meaningless.

Conclusion

The large number of trees examined in this study enables us to make some general observations about the relationship between size – the girth of a tree – and age: observations perhaps unsurprising to the trained dendrologist, but possibly more novel to readers whose main background is in history or archaeology. The first is, quite simply, that some very large trees are not in fact very old at all while, conversely, some rather unremarkable looking trees might be ancient.

Unsurprisingly (and as Mitchell, White and others have always emphasised) there are considerable variations in the rate at which different species put on girth. A hornbeam with a circumference of 3m could be a very old tree, perhaps planted in the early eighteenth century. An oak with this kind of girth would usually be of nineteenth-century date but could have been planted after *c.*1900, while a horse chestnut of this size would very probably post-date 1900. While we can make allowances for such variations, more problematic is the way that trees of the *same* species growing in different circumstances can put on girth at remarkably different rates. Even trees growing in close proximity, and in identical conditions of soil, aspect and drainage, can nevertheless display marked variations in girth, the consequence in part of irrecoverable idiosyncracies of life history – essentially, whether or not a particular tree has 'got away' and shaded out its neighbours. Such an observation is, again, not new (Lennon 2009, 170), but is perhaps insufficiently emphasised by many researchers. The conditions in which an individual tree grew during the earlier phases of its life (whether in an open situation or crowded by neighbours) are particularly important but such information is, in most cases, irrecoverable. While we can – with some caution – estimate the largest size to which trees of a particular age and species are likely to grow, many much smaller trees can thus be of equivalent age. In short, it is not possible to date an individual tree with any degree of confidence from its size alone.

This said, a second observation can, with some caution, be made on the basis of the above analysis: that there are few really old trees in the Norfolk landscape. There are no surviving examples of cedar, London plane, horse chestnut, willow, Scots pine or sycamore in the county which appear to have been established before *c.*1700; and probably no more than a handful of ash, hornbeam, maple, black poplar, beech, lime and sweet chestnut that pre-date *c.*1650. There are, it is true, large numbers of oaks which are older than this, but relatively few are likely to pre-date *c.*1500. Our legacy of ancient trees in the countryside is, in large measure, an early modern rather than a medieval one.

CHAPTER THREE

Trees of Farmland and Hedgerow

The dominance of oak pollards

The overwhelming majority of old and traditionally managed trees in Norfolk are found not in parks, woods or old wood-pastures but on farmland, principally in hedges: and, of these, the overwhelming majority are oaks (Figure 24). As already noted, these are mainly pedunculate oak (*Quercus robur*), although sessile oak (*Quercus petraea*) is thinly scattered in older hedges, especially along roadsides, right across north Norfolk from Trunch in the east to Swanton Novers and Glandford in the west, and as far south as Hevingham, Marsham and Ringland, although with particular concentrations around Edgefield and Baconsthorpe. This is the same general area in which *petraea* is well represented in ancient woods (Rogers 1984). Old and traditionally managed trees of other species are present in East Anglian hedges in very much smaller quantities: ash is the most important, followed by hornbeam and maple, and with only negligible numbers of other species, such as willow, black poplar and elm. But even ash is outnumbered by oak by a ratio of around twenty to one.

This massive dominance of oak is one striking feature of our old farmland trees. Another is that the overwhelming majority, in Norfolk as elsewhere in England, are former pollards rather than standards. In all, no less than 85 per cent of the trees recorded in fields and hedges with girths of 4m or more have been pollarded in the past; the figure for trees with girths greater than 5m rises to 90 per cent. These two key aspects of our heritage of old farmland trees – the dominance of oak, and the dominance of pollards – both raise difficult questions. Do they reflect the real balance of species and management practices in the pre-industrial landscape, and especially in that of the sixteenth, seventeenth and eighteenth centuries, when most of these trees were first established? And, if so, why were most hedgerow trees at this time oak pollards? Or is the dominance of oaks and pollards the consequence in some way of later patterns of survival and destruction?

It is sometimes suggested that most very old trees are pollards because this form of management has served to prolong their life (Read 2008, 251). A number of reasons for this have been suggested, most plausibly (and most clearly) by Lennon:

> Because the tree is regularly being cut back and the crown is constantly having to reform, pollarding can delay the emergence of the tree from the formative growth period. Where trees are continually pollarded the ring width will remain trapped in the formative cycle. This can extend the natural lifespan of the tree significantly and some of the oldest and largest trees in the country have been managed under this system for centuries. (Lennon 2009, 173)

It is also possible that trees being managed in this way were less likely to be blown over in storms than standards because of their lower-growing crowns. There may be some truth in these suggestions. But the dominance of pollards, in Norfolk at least, is largely the consequence of two rather more straightforward factors.

Firstly, trees in farmland (as opposed to those growing in ornamental parks and gardens) were primarily regarded in economic terms, as sources of wood and timber. A standard timber tree would not usually be allowed to reach any great age. In most cases, oaks of between 80 and 120 years were what were required by timber merchants. Moreover, at this kind of age the tree's rate of growth begins to slow, so that it made more sense to have it down and replace it with another. Only rarely would maiden trees be allowed to continue to grow into senescence, in places where they were valued as shelter for stock or for their appearance. It is noticeable that large maiden oaks with girths of more than 5m are far more common in landscape parks, gardens and churchyards than in the working countryside: 61 per cent of recorded examples are from the former locations, as compared with 34 per cent from the latter (the remainder are found in woods and relict wood-pastures). The scale on which timber trees were removed in the past from farmland and hedgerows is evident from even a cursory perusal of eighteenth- and nineteenth-century newspapers. And timber was not only cut down on a piecemeal basis, as and when it reached maturity: advertisements show clearly that trees were often stripped *en masse* from estates and farms, presumably when owners had a particular need of ready cash. In April 1835, for example, a sale was advertised in the *Norfolk Chronicle* of timber at Hethersett: 'about Two Hundred Oak, of various sizes, as they stand in the Hedge Rows'. Standard trees were thus unlikely to reach any great age. Pollards, in contrast,

FIGURE 24. An oak pollard in a hedge at Hedenham in south-east Norfolk. The vast majority of old and traditionally managed trees in Norfolk farmland are pedunculate oak (*Quercus robur*).

continued to have an economic importance late in life, and although their productivity might decline in senescence, there was less incentive to take them down as their short and often hollow boles made them of little use as timber. One reason why most old trees are pollards, in other words, is because only pollards have usually been allowed to reach a significant age.

Secondly, and perhaps more importantly, documentary and cartographic sources suggest that the majority of farmland trees, of all species, were in fact managed as pollards in the period before the nineteenth century. On a farm in Beeston near Mileham surveyed by Henry Keymer in 1761, for example, there were 413 pollards but only 104 timber trees: that is, 80 per cent of the trees were pollards (NRO WIS 138, 166X3) (Figure 25). Timber surveys from elsewhere suggest similar ratios: on a farm at Stanfield in *c.*1798 there were 192 pollards to 66 timber trees (74 per cent pollards): on a farm near Whissonset at the same time, 131 pollards and 59 timber trees (70 per cent pollards); on the same farm a few years earlier, 309 to 133 (again, 70 per cent); and just over the county boundary in Suffolk, at Thorndon, in 1742, 80 per cent of the trees present were pollards (NRO PD 703/45–6). Occasionally the proportion was

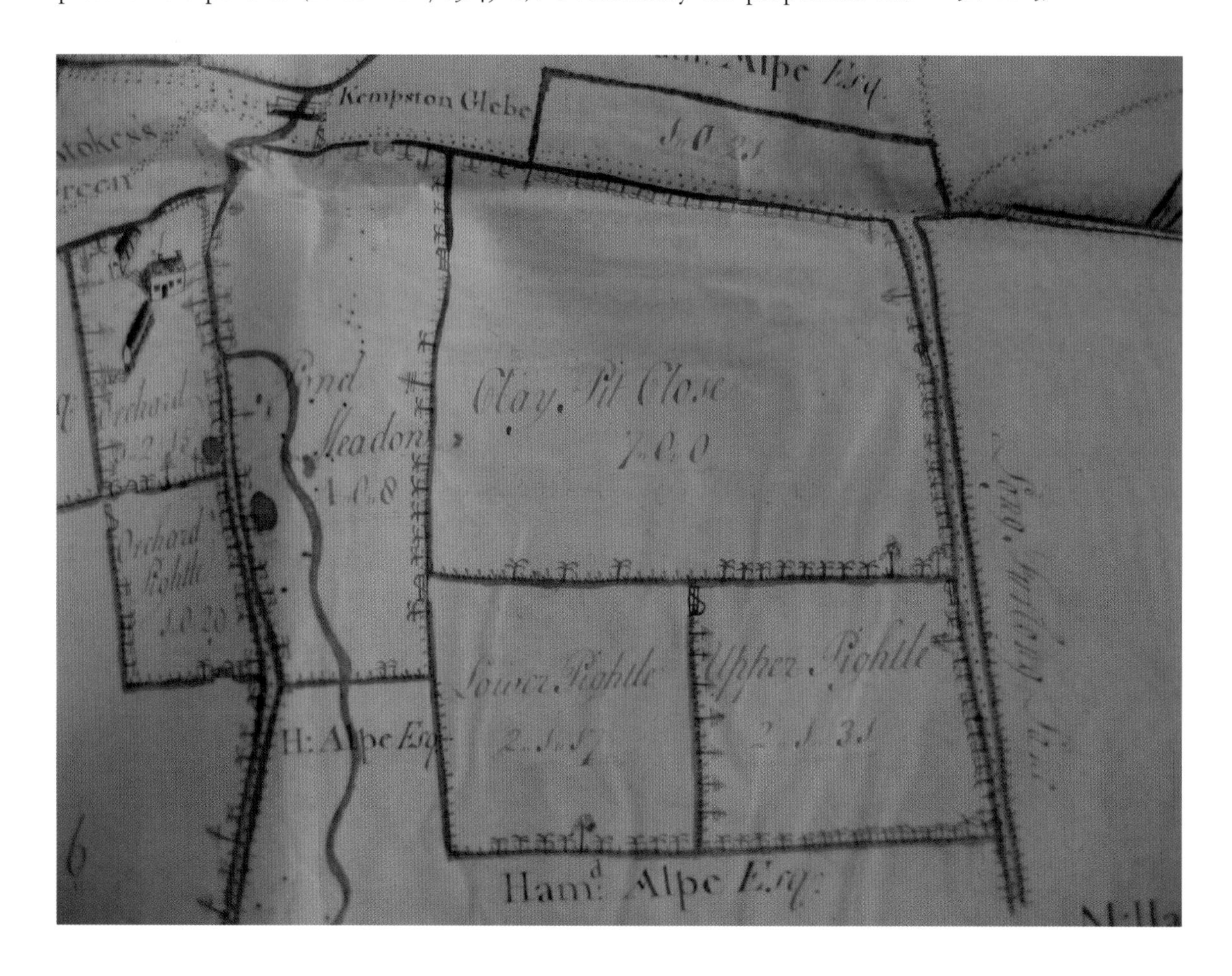

FIGURE 25. Hedgerow trees on a farm at Beeston-Next-Mileham in 1761. The surveyor, Henry Keymer, has carefully distinguished by symbol oak, ash, elm and 'pollards'. (NRO WIS 138, 166X3)

even higher, apparently reaching 100 per cent on a property at Ickburgh in 1651 (NRO Walsingham XXVI/4, 414X6). This dominance of pollards mainly reflects the greater need for relatively small pieces of wood, as opposed to large timbers, on the part of the local population. But it may also, in part, be a consequence of the importance of freeholders, and the independence and strength of tenants, on the heavier and more fertile soils of the county – the early-enclosed areas where a high proportion of such trees are to be found. Even where large estates held land in these regions it was often in the form of splintered parcels rather than unitary blocks, so that their tenants escaped close supervision. Both medieval custom and the terms of post-medieval leases generally stipulated that the tenant had the right to take wood but not the timber from the farm, and the temptation to convert any young tree into a pollard was thus overwhelming. Either way, the early modern landscape was dominated by pollarded trees and for this reason alone it is hardly surprising that most old trees are former pollards.

The question of how far the dominance of oak as a hedgerow veteran tree reflects its real pre-eminence in the countryside of Norfolk in the sixteenth, seventeenth and eighteenth centuries is more difficult to answer. The fact that oak so greatly outnumbers ash is particularly surprising, given that ash wood, being tough and flexible, had a host of uses. It was employed to make tool handles and parts of ploughs, carts and wagons (especially the wheels), as well as for hurdles and fencing (Evelyn 1664, 22–3). It could be used as fodder for cattle. Above all, it makes excellent firewood, burning well even when green: in Evelyn's words 'the sweetest of our forest fuelling, and the fittest for Ladies Chambers' (Evelyn 1664, 23). In one respect, of course, the present balance of old trees in the Norfolk countryside is the consequence of a relatively recent development: the wholesale loss of elm from the landscape in the 1960s and 1970s brought about by Dutch elm disease (prior to which elm made up around a quarter of farmland trees in Norfolk (Norfolk County Council 1975)). Although elm remains common as a hedgerow shrub, as a full-grown tree it is now reduced to a small number of examples, very few of which are of 'veteran status' (of the fifteen examples recorded in the various surveys on which this volume is based, none has a girth of more than 5m). This in turn raises a wider possibility – that the clear preponderance of old oaks over other trees is a consequence of survival, of that species' longevity compared with others. While oaks can live for 700 or more years, ash trees seldom thrive after around 250 years, and relatively few survive beyond 300 (Thomas 2000, 257). The essentially economic demands made on farmland trees will have accentuated this natural difference, especially in the case of pollards. Once in decline, ash trees managed in this way were probably rapidly felled by landowners. Discussing ash, Moses Cook advised that 'If once you find your Pollard grows much hollow at the Head, down with it as soon as may be' (Cook 1676, 78), advice that was repeated by Evelyn in the 1669 edition of his famous book *Sylva* (Evelyn 1669, 43).

This said, documentary evidence does suggest that while ash (and elm) were more common as hedge trees in the early modern period than the numbers of old specimens now surviving suggest, oak was nevertheless dominant, although to a lesser extent and one that probably varied from area to area. At Langley in the south-east of the county no less than 70 per cent of the trees recorded in a mid-seventeenth-century estate survey were oak and only 30 per cent ash (NRO NRS 11126). But at Buckenham near Blofield in the 1690s 49 per cent were oak and 44 per cent ash, together with 6.6 per cent elm (as well as six poplars and 'young' trees of unspecified species) (NRO Beauchamp-Proctor 334; NRO NRS 11126). On the farm at Beeston surveyed by Keymer in 1761 (NRO WIS 138, 166X3) there were actually more ash than oak trees recorded in the fields and hedges – 47 per cent and 41 per cent respectively, with 12 per cent elm – but the map does not tell us the species of the pollards which, as we have seen, massively outnumbered the timber trees on the farm, and it is quite possible that the majority of these were oaks, given the fact that where old trees survive today in any numbers in the locality (as, for example, at nearby Brisley) pollarded oaks are found closely planted in many hedges. Diocesan timber accounts from the late eighteenth century, recording timber felled and sold at Hindolveston, North Elmham, Arminghall and elsewhere in the county, record a considerable preponderance of oak over ash and elm, although the accounts are unclear in places and some of the trees may have been from woodland as well as farmland (NRO DCN 22/10). In contrast, sales of hedgerow timber advertised in the local papers in the eighteenth century often suggest a more even balance of oak and ash. In February 1776, for example, an advert appeared in the *Norwich Chronicle* announcing the sale of 'forty five oak, thirty ash and eight elm' growing on a farm at Tuttington near Aylsham. On balance, oak probably was the most numerous tree in the Norfolk countryside in the early modern period, although not by as great a margin as the relative numbers of old trees in the landscape today would suggest, its representation having been increased by its longevity relative to ash and by the wholesale loss of elm.

The management of hedgerow oaks

Oak is only moderately common as a shrub in Norfolk hedges. It appears to have been but rarely planted as a hedging plant, is a slow colonist, and where it does occur in hedges it is often clearly regenerating from the stumps of long-felled trees. Most oaks found in hedges were thus probably deliberately planted there, rather than representing adventitious plants which have been allowed to grow to maturity. This is particularly true of specimens found in the species-poor, hawthorn-dominated hedges of the eighteenth and nineteenth centuries, but even in older, mixed hedges the frequency of oak as a tree is considerably greater than its frequency as a shrub.

Many hedgerow oaks in Norfolk (around 60 per cent) were pollarded at between 1.5m and 2.5m above the ground. A tiny minority were cut lower than this, and around 40 per cent higher, although only *c.*3.5 per cent of the total at a height greater than 3.5m. There are no obvious practical explanations for these variations. It might be thought that pollarding was carried out at a higher level where, for example, trees grew in hedges beside roads and lanes, in order not to obstruct laden carts, but specimens in such locations seem to display a range of pollarding heights no different to those of trees growing in other boundaries. Particularly perplexing is the way in which adjacent pollards in the same hedgerow, in some cases probably planted at the same time, can display widely differing pollarding heights. For the most part, pollarding height is probably arbitrary and random, simply reflecting the choice made at first cutting – probably when the tree was about ten to fifteen years old – which was presumably influenced by such practical considerations as the height of the vegetation in the surrounding hedges, and the growth pattern of the tree in question. Some specimens display two or more levels of cutting, perhaps where pollarding lapsed for a while and was then resumed, and many have a wide calloused damage layer, extending over a metre or more, where the level of cropping has gradually migrated upwards.

A more puzzling question, but one seldom asked, is why hedgerow oaks were pollarded at all. On the face of it, the answer is straightforward. Pollarding was, in effect, aerial coppicing, a method of producing wood out of reach of grazing livestock. Coppicing was practised in woods, where stock were excluded by fences, banks and ditches. Pollarding was employed in wood-pastures, parks and the wider countryside, where a variety of stock was grazed around or beneath the trees for at least some of the time. The problem with this argument, in so far as it applies to hedgerow pollards, is that agricultural writers and estate accounts make it clear that, at least by the later eighteenth century, most hedges in Norfolk were themselves managed by coppicing. They were cut back to the ground, like linear woods, at intervals of ten to fifteen years, rather than being managed by plashing or laying, the more complicated process, common in Midland and western counties, by which the hedge was less drastically cut back and its principal elements cut two-thirds of the way through at the base, bent over and woven through regularly spaced upright stakes, so that it came to form a dense, impenetrable wall of vegetation. William Marshall thus described in the 1780s how, in the area of north-east Norfolk with which he was most familiar, plashing was entirely unknown: coppicing was preferred because it produced more usable wood. Indeed, he asserted that the entire supply of wood and timber in the district 'may be said, with little latitude, to be from hedge-rows' (Marshall 1787, 96). Coppicing, which was carried out in the winter months, demanded less skill than laying, but required some thought over farm management as the new growth needed to be protected from browsing livestock for several years. Where, as was often the case on the claylands in the centre and south of the county, deep ditches accompanied the hedge, animals were simply excluded from the unditched side for the necessary period of time. Alternatively, or in addition, the hedge might be temporarily protected with hurdles, or with lines of staked brushwood – thorns and other material cut from the hedge. Plashing is only mentioned by eighteenth- and nineteenth-century writers as a method of dealing with neglected or decayed hedges, many of which were by this stage composed entirely of thorn and managed simply by regular trimming. R. N. Bacon, writing in 1844, thus quoted with approval the comments of one Mr Withers, from Holt: 'Where hedges are properly formed and kept, they can seldom require to be plashed; but this mode of treating a hedge is most valuable in the cases of the fences abounding with hedge-row timber, when from neglect or any other cause the hedge has become of irregular growth' (Bacon 1844, 390).

If hedges were being managed by coppicing it is not immediately obvious why the oaks growing within them should have been managed any differently – why, that is, they could not have been cut down to ground level, on rotation, with the other shrubs. There are several possible explanations. Firstly, there is some evidence that the ways in which hedgerows were managed in Norfolk changed over time, and that in the period before the eighteenth century, when many of our old oaks were first pollarded, laying and plashing were more commonly practised than in Marshall's day. Certainly, there are a number of possible references to this form of management in early estate accounts. At Raynham in west Norfolk in 1664, for example, the steward gave instructions 'to plash the hedges behind the garden and that next the heath' (Saunders 1917), while in 1516 John Skayman, an earlier steward for the Raynham estate, recorded how he 'was at Bermer [Barmer] to see who [how] the stakes and hogyn wood lay', ready for the hedgers there. The latter promised him that as soon as the weather improved they 'schuld be heggyn ther as fast as thei cane for they cane done no good now for by cause the frost is so grete that thei canne sette no stakes' (Moreton and Rutledge 1997, 121). The 'stakes' were presumably the upright poles (called 'stabbers') through which the bent-over stems (or 'pleachers') of a laid hedge were woven. Similarly, in 1472 payments were made for 'digging, plashing and hedging' the boundary of Sporle Wood (Gairdner 1895). Plashing may thus once have been normal practice in medieval and early modern times, gradually declining for a variety of reasons in the course of the eighteenth century. However, a substantial proportion of oak pollards were unquestionably planted after *c.*1700 (many occur in ruler-straight, species-poor hedges), and were thus first established, and managed, within the period in which – to judge from the available evidence – coppicing and cutting were the almost universal methods of hedgerow management.

A more likely explanation is that oaks were cut on a longer cycle than the hedges in which they grew. This may have been to produce a different range of produce than the relatively thin wood which would be derived from coppicing hedgerow shrubs on a ten-year cycle – material with a larger diameter, suitable for good-quality fuelwood, fencing and minor building components, rather than for faggots and oven firing. In addition, oak grows comparatively slowly and might well be out-competed by other species (especially hazel and ash, common components of Norfolk hedges) if it was simply coppiced with the other shrubs. It is also possible that some of the older hedgerow oaks were cut at a different time of the year to the hedge in which they grow, in the summer as opposed to the winter, because they were being managed to produce 'leafy hay' – winter fodder for cattle – a practice recorded even in relatively recent times elsewhere in England, and which still goes on in some parts of Europe (Fleming 1998; Halstead 1996; Slotte 2001; Spray 1981; Read 2008). Such trees, to judge from modern Scandinavian parallels, would also have been cut on a *shorter* rotation than the adjacent hedge, generally every two to six years, 'before the amount of woody material exceeded the amount of leaves' (Read 2008, 251). As we have already noted (above, p. 5), some trees appear to have been managed primarily with browse or fodder in mind. A survey of the manor of Redgrave, just over the county boundary in Suffolk, made shortly after the dissolution of the monasteries, typically described how on the 'seyd mannor and dyvers tenementes there … be growing 1,100 okes of 60, 80 and 100 yeares growth part tymber parte usually cropped and *shred*' (WSRO Accn 1066; our emphasis). Shredding involved shaving off the side branches of the tree so that it grew to resemble a kind of fuzzy bottle brush, producing an abundance of fresh, leafy growth. Oak was the most commonly shredded species, although elm, poplar and ash were also managed in this way. A number of ancient trees in Norfolk, including former hedgerow trees of probable medieval date now growing within the parks at Kimberley and Houghton, may be surviving examples of shreds. They do not have the well-defined damaged layer and spreading branches characteristic of pollards; but nor do they look like normal standard trees, in that they have branches extending rather evenly down the length of the trunk, which itself displays extensive areas of calluses and much epicormic growth.

Not all farmland oaks grew in hedges. In the old-enclosed claylands, in particular, seventeenth- and eighteenth-century documents suggest that it was also customary to group pollards and 'shreds' along the field margins, in 'rows' three or four trees deep. This was one aspect of a more general phenomenon in these landscapes. Where fields were used as arable the cropped area frequently did not run right up to the boundary hedge; there was often a 'hedge green', an uncultivated strip, on which the ploughteam could turn, and which was mown for hay or used for grazing tethered cattle. Such 'rows', in effect diminutive linear wood-pastures, are shown on many early maps but were systematically removed in the eighteenth and nineteenth centuries as landowners grew more hostile to pollarding and as the area under arable, as opposed to permanent pasture, increased (Wade Martins and Williamson 1999, 21–7, 67–9). Only a handful of old pollards appear to represent survivors from such features.

Other hedgerow species

Ash is the second most common of the ancient and traditionally managed trees found on Norfolk farmland. It occurs mainly as a pollard but occasionally as a standard. As noted, its wood had a range of uses, although it was valued above all as fuel. As Moses Cook put it, 'Of all the wood that I know, there is none burns so well green, as the Ash' (Cook 1676, 76). Ash leaves and twigs are also avidly consumed by sheep and cattle, and dry well to form leafy hay. Although, as we have seen, there may have been more ash trees in fields and hedges in the past, perhaps approaching the numbers of oak in some parts of the county, they were probably always less common than oaks.

There are a number of possible reasons for this. Firstly, a number of early writers cautioned against planting ash standards and, in particular, pollards in field boundaries on the grounds that they robbed the adjacent soil of nutrients and water: in the words of Moses Cook, 'the Summer after a tree is lopped, it shall rob the Corn more than another bigger tree standing by it, as may be visible by the growth of the Corn' (Cook 1676, 76). In the mid-nineteenth century Brown similarly warned that 'the roots of the ash are of all other trees the most searching upon land, and impoverish it very much'. He also noted that it shaded the crop more than most trees: 'when standing alone and exposed to free air, its top is extremely liable to diverge off into large limbs' (Brown 1861, 397). Secondly, ash may have been less common as a hedgerow tree because it was particularly abundant as a hedgerow shrub, today occurring in around 80 per cent of Norfolk hedges. It appears to have been a popular choice as a hedging plant in the period before the mid-eighteenth century when mixed hedges were widely planted in the county. It also colonises quickly and maintains itself well in hedges, especially ones managed by coppicing (indeed, its success as a rapid coloniser also ensures that it has now widely established itself even in hedges of late-eighteenth- and nineteenth-century date that were originally planted solely with thorn) (Barnes and Williamson 2006, 81). Traditionally, ash thus provided the main source of fuelwood in coppiced hedges and, given its rapid growth, even a hedge managed on a relatively short rotation would produce an abundance of sizeable ash poles. There may thus have been less incentive to manage ash as a pollard within hedges, at least as a source of fuel. In this context it may be significant that a large proportion of known ash pollards are found in the pure thorn hedges planted in the eighteenth and nineteenth centuries (Figure 26). These were not usually managed by coppicing, but by regular trimming, and they contained no useful wood. In such a context, pollarding ash trees for firewood made obvious sense. Where ash was pollarded in older, mixed hedges it may, as in the case of oak, have been because it was being cut at a different time to the rest of the hedge, and/or on a different rotation: perhaps for fodder, or to provide larger poles for specialised uses. Either way, pollarding heights appear similar to those for oak, with small numbers (3 per cent) cut at less than 1.5m; most (61 per cent) at between 1.5m and 2.5m; around 43 per cent at more than 2.5m; and only a small number (3 per cent of the total) above 3.5m.

As well as being far less common in hedges and on farmland than oaks, old ash trees are also less widely distributed. Whereas ancient and traditionally managed trees of the former species can be found in most situations, old ash specimens are rarely found on dry, acid soils, clustering instead in calcareous

FIGURE 26. Ash pollards growing in hedges planted when the commons to the south of Wymondham were enclosed in 1816. Pollarded ash trees are often found in relatively young hedges.

and, in particular, damp locations. Even on the claylands in the south of the county there is a clear tendency for specimens to be found on the damp plateau soils of the Beccles Association or beside streams or on flood plains rather than on the drier, sandier clays of the Burlingham Association, which occupy the gently sloping sides of the valleys cutting through the boulder-clay plateau (Figure 27). In contrast, *young* ash trees can be found growing on all the principal soils types in the county, and hedges containing ash as a shrub are likewise distributed fairly evenly. Evidently, dry and acid conditions militate against *trees* of this species achieving any significant age.

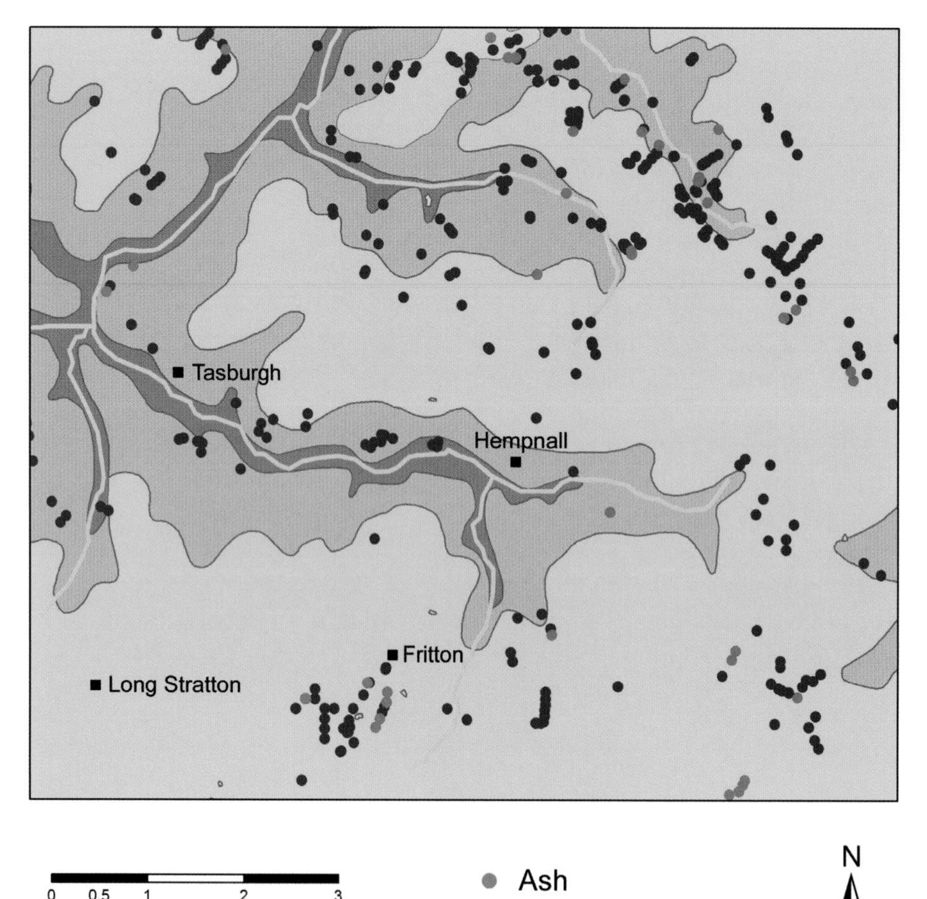

FIGURE 27. The distribution of old and traditionally managed ash and oak trees on the south Norfolk claylands. Note how old ash trees tend to be found in damper locations, either close to watercourses or on the damp Beccles Association soils (light green) occupying the level clay plateau. They are much rarer on the lighter clay soils of the Burlingham Association (light brown), which occupy the sloping sides of the principal valleys.

Hornbeam is the third most common veteran tree in the Norfolk countryside, with seventy examples recorded in hedges (or former hedges) and as free-standing specimens in fields (Figure 28). It was perhaps never a common tree: it hardly figures in seventeenth- and eighteenth-century timber surveys. Almost all recorded examples are pollards, as is the case in other parts of England. As Loudon noted in 1838, 'the few old trees remaining in this country, of any size, are pollards' (Loudon 1838, III, 205). Their distribution within Norfolk is, even more than with ash, strongly correlated with the damper soils. Around a half are found on a single soil type – the heavy clays of the Beccles Association, which not only characterise the level boulder-clay plateau in the centre and south of the county but also occur as an isolated pocket in the far west of the county, on the edge of the Fens. Most other examples are recorded in low-lying positions beside streams or on the flood plains of rivers, especially those draining the clays (Figure 29). Once again, few examples occur on the sandier clay soils, those of the Burlingham Associations, and virtually none on more freely draining and acidic soils. It is possible that, like oak, hornbeam was deliberately planted as a hedgerow tree. But most, if not all, examples, occur in old, species-rich hedges and probably began life as hedgerow shrubs which grew into young trees during periods of neglect. Hornbeam tends to be found as a shrub only in the oldest, most species-rich hedges, and almost always on heavy clay soils. Indeed, the distributions of hedges containing hornbeam as a shrub and of hornbeam pollards are strikingly similar, and most of the hedges which contain hornbeam pollards also seem to contain hornbeam as a shrub.

The white, close-grained wood of hornbeam is the hardest produced by any British tree. In Edlin's words, it is 'so hard and difficult to work that it is no longer used on any scale in commercial forestry' (Edlin 1947, 12). But in earlier centuries it was a valued commodity for such things as butchers' blocks, mallets, screws and, in particular, pulleys and cogs in mills, for which reason it was sometimes called the 'engineer's timber'. Evelyn described how it was used for 'Millcogs &c. (for which it excels either Yew or Crab), Yoak-timber ... Heads of Beetles, Stocks and handles of Tools (for all which purposes its extream toughnesse commends it to the Husbandman)' (Evelyn 1664, 29). It is probable that the branches cut from hornbeam pollards were employed in such specialised ways, and that the pollards themselves were managed on a

FIGURE 28. Hornbeam pollard in a hedge at Great Melton. Pollarded hornbeams are a rare but distinctive feature of the old-enclosed clayland landscapes of southern and central Norfolk.

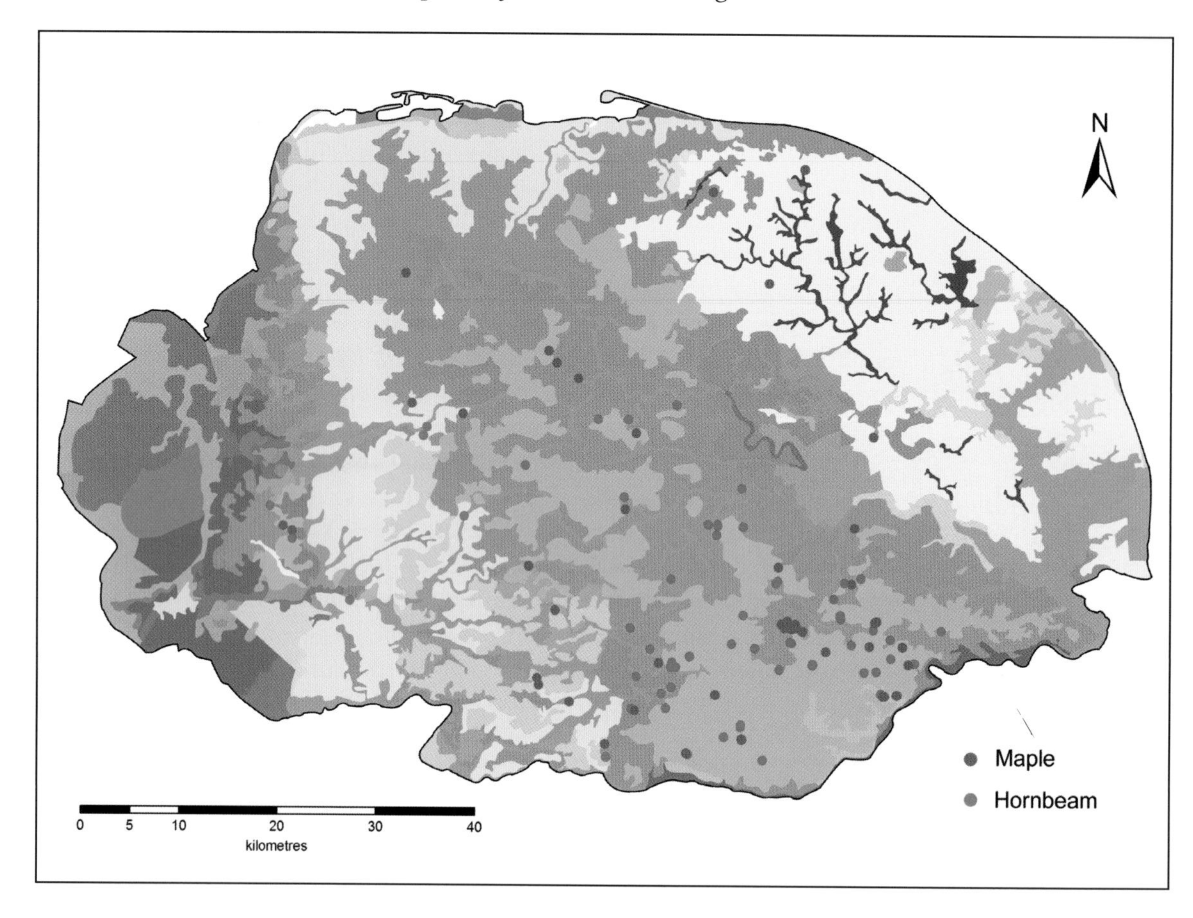

FIGURE 29. The distribution of old and traditionally managed hornbeam and maple, and principal soils types, in Norfolk. Key: see Figure 2.

longer rotation than the adjacent hedge. But the principal use for hornbeam poles, from both pollards and coppices, was as fuel wood. In Loudon's words, 'As Fuel, the wood of the hornbeam should be placed in the highest rank … as it lights easily, and makes a bright flame, which continues a long time, and gives out abundance of heat' (Loudon 1838, III, 2009). Evelyn similarly commented that hornbeam 'made good firewood, where it burns like a candle' (Evelyn 1699, 64). In addition, Loudon described how the charcoal was 'highly esteemed', and noted how the leaves were dried in France and used for fodder. Similar uses might have been made of the tree and its wood in medieval and early post-medieval Norfolk. As already noted, few hornbeam pollards attain any significant size, in spite of their probable antiquity. The height of cutting exhibits a similar range to ash and oak, with most examples (56 per cent) being polled at between 1.5m and 2.5m, 39 per cent higher than this, and only 5 per cent above 3.5m.

Old maple trees are rarer in Norfolk than old hornbeams, with only around fifty examples recorded, all from hedges or former hedges. They have a slightly wider distribution, being found both on the heavier plateau clay soils of the Beccles Association and on some of the light calcareous loams in the west of

FIGURE 30. Maple pollard at Burston. Maples appear to have been pollarded at a significantly lower height than other hedgerow trees.

the county. Of the maidens and pollards (as opposed to hedge stools), all but three of the recorded examples have girths of less than 3m, and only one is greater than 3.5m in circumference. In general, pollards of this species have been cropped at a noticeably lower height than oak, ash or hornbeam, with 43 per cent at less than 1.5m, and none higher than 2.5m (Figure 30): very low pollards like this were often described in the seventeenth and eighteenth centuries as 'stubs'. Maple wood makes reasonable fuel, but it was also traditionally valued for its hard, fine-grained character. In Evelyn's words:

> The Timber is far superior to Beech for all the uses of the Turner, who seeks it for Dishes, Cups, Trays, Trenchers &c., as the Joyner for Tables, Inlaying, and for the delicateness of the grain, when the knurs and nodosities are rarely diapred, which does much advance its price. (Evelyn 1664, 28)

In spite of such advantages, maple was rarely planted as a timber tree or managed as a pollard, but was instead usually grown in woods as part of the coppice understorey. Indeed, Moses Cook – although an enthusiastic supporter of the maple as underwood – advised against growing it in hedges on the grounds that:

> It receives a clammy Honey-dew on its Leaves, which when it is washed off by the Rains, and falls upon the Buds of those Trees under it, its Clamminess keeps those Buds from opening, and so by degrees kills all the Wood under it: therefore suffer not high Trees or Pollards to grow in your hedges, but fell them close to the Ground, and so it will thicken your hedge, and not spoil its Neighbours so much. (Cook 1676, 99)

Whether such an opinion was widely shared is unclear, but it is unlikely that many of the examples found in Norfolk were intentionally planted with either timber or eventual pollarding in mind. Most are found in mixed, species-rich hedges and, as with hornbeam, it is likely that they were originally planted as part of the hedge, or self-seeded there subsequently, and were later pollarded in an opportunistic fashion – perhaps when an outgrown hedge was being brought back into management.

Elm was widespread as a hedgerow tree in some parts of Norfolk before the onset of Dutch elm disease in the 1960s. Fortunately, it was never as common here as oak, and the impact of the disease was thus much less than in some neighbouring areas, such as south Essex, where it was the dominant farmland tree (Wilkinson 1978). Elm is now very rare everywhere in England as a tree, although common enough as a hedgerow shrub: as already noted, only when it begins to grow into a true tree, with fully formed bark, does it fall victim to the disease. So long as hedges are kept managed by frequent cutting the effects of the disease are minimal. Most hedgerow and farmland elms in Norfolk are of a particular variety, the East Anglian elm (*Ulmus minor* or *carpinifolia*) (Rackham 1986a, 232–47; Richens 1983). This has a longer, more pointed leaf and a more vigorously suckering habit than the English elm (*Ulmus procera*), which dominates the landscape of the areas immediately to the south and west of East Anglia – south Essex, central and western Hertfordshire and the Midlands. The East Anglian elm does not normally reproduce from seed, but by suckering, and the botanist R. H. Richens has mapped the extraordinarily restricted distribution of particular clones (in Mabey's words, 'there are now almost as many different kinds of elm as places in which they grow' (Mabey 1998, 100)). Richens also attempted – rather less convincingly – to relate these to successive waves of prehistoric and early historic invaders. In fact, pollen analysis shows that *U. minor* was present in Britain before the arrival of the first farmers, and while some of its many varieties may have arisen naturally over the last few thousand years, others probably pre-date the return of the tree to Britain after the end of the last glaciation (Rackham 1986a).

Elm had a myriad of uses. It was employed, although much less than oak, in the construction of timber-framed buildings (in Denton, in the south-east of the county, some medieval buildings are built almost entirely of elm). It was more frequently used for floorboards, doors and panelling. It was also the main wood used for weatherboarding, mangers, bins in granaries – and for coffins. It was, above all, used in contexts where it would remain wet for much of the time – for mill sluices, keels for ships – and elm trunks were hollowed out to make drainage pipes. Although small-diameter material was required for some of these specialised uses (such as piling), most required sizeable limbs or trunks. Elm is a common component of mixed hedges in the county and where trees were managed by pollarding, rather than being coppiced with the rest of the hedge, it may have been to allow the production of comparatively large material to be employed in some of these ways. Elm makes indifferent firewood, unless allowed to season for a long period of time. Its foliage was, however, widely employed as fodder for cattle throughout Europe, and some Norfolk pollards may likewise have been cut on a short rotation to supply 'leafy hay'. Maps and estate surveys suggest that elm never accounted for more than 15 per cent of hedgerow trees in Norfolk, but these sources are relatively few in number and it is possible that it was more common than this in the past, at least in parts of the claylands. It is often mentioned, in passing, in other kinds of documentary source. Randall Burroughes thus recorded in his farming journal for January 1798 how his men had been employed 'sawing off the crown and stubb of a pollarded elm against the direction post' (Wade Martins and Williamson 1996,

94). And elm, as noted, is widespread as a shrub in Norfolk hedges, especially on the clays (where it is present in well over half of examples) but also in the oldest hedges on the light loams in the north-east of the county. In some cases it may have been planted as a hedging plant (an extent of Snetterton from 1576 refers to an elm hedge (Davison 1973, 349)); while in some of the oldest hedges it may have suckered in from adjoining areas of woodland at an early stage in the process of clearance and colonisation. But, in most cases, hedgerow shrubs have probably suckered from long-lost hedgerow trees, either standards or pollards. Indeed, in the claylands elm is even well represented in (that is, present in around 50 per cent of) the straight, hawthorn-dominated hedges planted in the later eighteenth and nineteenth centuries, especially when areas of common land were enclosed (Barnes and Williamson 2006, 76). However common elm may or may not have been in the past, it is not common now. Dutch elm disease has taken a terrible toll. Only thirteen living or recently deceased veterans of *Ulmus carpinifolia* have been recorded in Norfolk, most of which are former pollards. None have girths greater than 5m, but they may be older than they look. In addition, two pollarded examples of wych elm (*Ulmus glabra*), a variety probably not native to the region but widely planted, were recorded at Hempnall in the south of the county (with girths of 3.0m and 3.2m).

A great deal has been written in recent years about the native black poplar (*Populus nigra* subsp. *betulifolia*), one of Britain's most attractive trees, as well as being one of its rarest (Cooper 2006) (Figure 31). There are now only around 7000 examples surviving in England, mainly to the south of a line drawn from the Mersey to the Wash, and with marked concentrations in Shropshire, Cheshire and the Vale of Aylesbury in Buckinghamshire. In Norfolk the tree is unquestionably rare, with just over a hundred known examples of all ages. Some of these are in cemeteries or gardens, but most grow in farmland, in meadows, pastures or hedges. Around half are pollards. The tree may have been more common in the county in the past, although it is noteworthy that the Norwich barrister Charles Ambler, writing in 1741, asserted that 'the black poplar rarely grows with us' (NRO DN/PCD6/5). Recorded examples are widely scattered but show a preference for well-watered sites. Several examples are recorded from low-lying flood plains in Breckland, at Northwold, Roudham, South Lopham, Sturston, Tottington, Weeting and Larling, forming a continuation of the concentration which has been noted in the adjoining parts of Suffolk (Cooper 2006, 45). There are scattered examples in the Wensum valley and in a line lying a little inland from the north coast, mainly in the lower reaches of the principal rivers with outfalls here, as at Binham and Wiveton. Yet although the natural habitat of the black poplar is usually considered to be flood plains, marshes or river banks (where it may once have formed extensive areas of flood-plain forest) most surviving specimens are in fact located on relatively high ground, on the damp clay soils in the south and centre of the county (usually those of the Beccles Association), where they often occur beside ponds (Figure 32). There are concentrations in and around the village of Thorpe Abbots and

FIGURE 31. A magnificent black poplar growing in a hedge beside a road in East Harling.

the town of Wymondham, the latter perhaps associated with the wood-turning industry which was traditionally based there (Rogers 1993, 378). A number of specimens are associated with greens and commons, or former greens and commons, as at Pristow Green, in Tibenham, or Wacton Common. Many are located beside farmyards, where they were perhaps planted for aesthetic reasons, but possibly for practical ones, as the wood was noted for its shock-absorbent qualities, which made it ideal for the construction of carts. The most striking association, however, is with a range of industrial sites – former kilns (thirteen examples in Norfolk of surviving or recorded trees lie within 200m of such a site); mills (twelve examples); smithies (six examples) and malthouses (four examples) (Barnes *et al.* 2009, 35–6). This association perhaps reflects the fire-

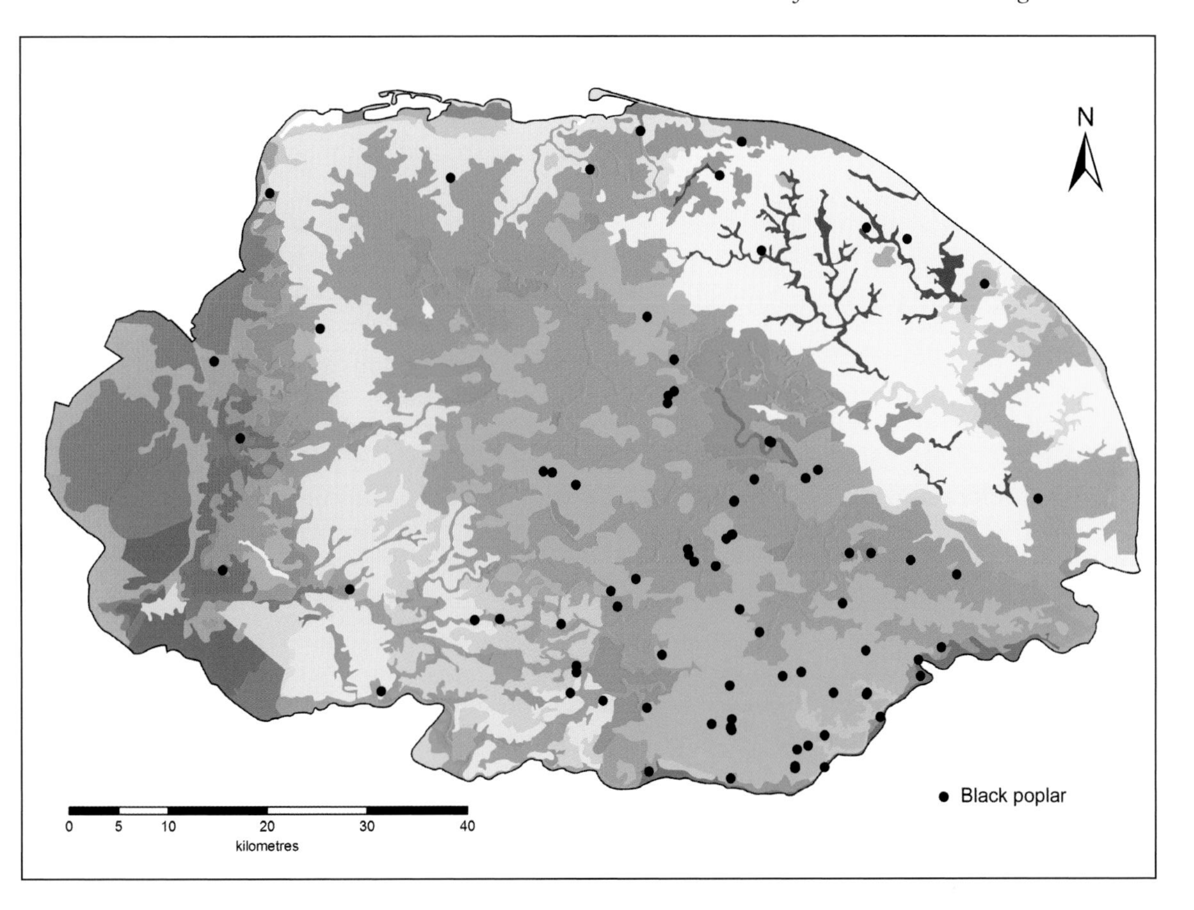

FIGURE 32. The distribution of black poplars, and principal soils types, in Norfolk. Key: see Figure 2.

resistant qualities of the wood: it was useful to have a handy source for repairing such things as bellows, mill machinery and malthouse floors. Charles Ambler described how poplar wood was particularly valued by the manufacturers of pumps and bellows (NRO DN/PCD6/5). Certainly, there are hints in the documentary record that eighteenth-century industrialists took a keen interest in the tree. The businessman Meadows Taylor, of Diss on the Norfolk–Suffolk border, purchased several meadows, damp pastures and low commons in the area (NRO MC 257) and in 1794 bought nineteen 'poplar timber trees' growing on The Lows Common in Palgrave in Suffolk, perhaps to make repairs to his malt houses and hop kilns across the river in Diss. It is noteworthy that one of the few examples noted by James Grigor in his *Eastern Arboretum* of 1841 was at Narborough, 'on the road-side in front of the mills of Mr Everett' (Grigor 1841, 343).

Willow and alder are often found on damp, low-lying ground, occasionally in hedges but more usually free-standing in pastures and meadows. We have very little evidence concerning the antiquity of such trees, although few surviving examples are probably of any significant age; nor about how they were used and managed in the past. Neither tree makes good firewood, but willow was used to make baskets, alder for scaffolding and to produce charcoal and also (because it is resistant to rot if kept wet) for piling and jetties. Most of these uses were, at least in post-medieval Norfolk, probably supplied from coppices and withy-beds, rather than from standards and pollards. As we shall see, willows were also planted to stabilise banks and roads, especially in the Broads.

Oak, ash, maple, hornbeam, elm, willow, alder and black poplar together account for almost all the farmland trees of any real antiquity found in the Norfolk countryside. In addition, a handful of examples of beech and sycamore are recorded as old hedge pollards, as are occasional examples of sweet chestnut (at Holkham and Felbrigg). Such deviations from normal practice are perhaps associated, in particular, with the home farms of large landed estates, rather than with the lands of small proprietors, and may have been more common in the past, at least by the nineteenth century, than is suggested by surviving trees in the countryside. In March 1868, for instance, the 'large quantity of superior Hedgerow Timber Trees' advertised for sale at Bacton in the *Norfolk Chronicle* included, beside 100 oaks and 100 ash, 150 'sycamore, elm, beech and spruce'.

The decline of pollarding

As we have seen, the eighteenth century saw increasing hostility towards the practice of pollarding on the part of large landowners and 'polite society' more generally. In part this was for aesthetic reasons – the enthusiasm for 'natural' forms in parks and gardens and the growing opposition to topiary were easily extended to the trees growing on neighbouring estate land (Thomas 1983, 220–21; Petit and Watkins 2003, 171). More importantly, pollards were increasingly associated with backward peasant agriculture, for large landowners obtained their firewood from plantations of coppices, and both they and their middle-class neighbours increasingly burnt coal in their grates. Keen agricultural improvers were especially hostile to pollards, for their low, dense heads spread a particularly deep pool of shade. As William Marshall put it: 'We declare ourselves enemies to Pollards; they are unsightly; they encumber and destroy the Hedge they stand in (especially those whose stems are short), and occupy spaces which might, in general, be better filled by timber trees; and, at present, it seems to be the prevailing fashion to clear them away' (Marshall 1796, 100–101).

There has been a tendency on the part of many researchers to take such comments at face value and to believe that the strictures of such men were widely influential. Petit and Watkins, for example, have described how 'local field and documentary evidence supports the idea of a decline in pollarding from the late eighteenth century onwards' (Petit and Watkins 2003, 171). Lennon similarly notes that pollarding 'has declined sharply over the last 200 years' and that 'the vast majority of the old pollards that we see today have not been actively managed in the traditional method for a long period of time, in some cases centuries' (Lennon 2009, 173). Although such statements are broadly true, the decline of pollarding should neither be hastened nor exaggerated. Even an enthusiastic late-eighteenth-century 'improver' such as Randall Burroughes of Wymondham, while he might happily fell old, outgrown oak pollards, was equally happy to crop ash pollards, thus recording in February 1795:

> Monday, Tuesday and part of Wednesday we finished lopping the ash trees beginning at common gate against Mr Hart & continued to the corner against the road. About four score and ten faggots were sent to Mrs Denton the remainder stacked at home, the round wood having been separated before tying and reserv'd for hurdle; the whole contents might be 300 faggots. (Wade Martins and Williamson 1996, 51)

Small farmers, in particular, continued to value their hedgerow trees as sources of wood, continued to crop existing pollards and to create new ones right through the eighteenth and nineteenth centuries. The hedges surrounding the large, rectilinear fields created by parliamentary enclosure in the north and west of the county, in the land of large estates, contain relatively few pollarded trees. But where small commons on the claylands were enclosed in the early years of the nineteenth century pollards were not infrequently planted in the new hedges – sometimes ash trees, such as those which are a feature of the former commons around Morley, Besthorpe and Wymondham (see Figure 26), and sometimes oaks. Pollards are also widely found in hedges newly planted or realigned elsewhere on the heavier soils of the county in this period. Indeed, more than 730 oak pollards with girths of between 2m and 3m are known from the county, out of total of 2160 recorded examples, suggesting that as many as a third of existing hedgerow pollards of this species may have been first planted and cropped in the period after the middle of the eighteenth century. Ash pollards are perhaps more striking in this respect for, to judge from their locations and girth measurements, as many as two-thirds of examples recorded in the county may have been planted and first cropped after *c.*1750.

New pollards thus continued to be created on some scale, at least in the south and east of the county, well into the nineteenth century. Indeed, pollarding remained common practice well into the twentieth: many older farmers and landowners in East Anglia today describe how particular trees were still being regularly cut within their own lifetime. Even the creation of entirely new pollards may have continued, albeit at a low and declining rate, into the last century. In this context, attention should be drawn to a small number of oak pollards recorded in hedges in the county – no more than fifty-six – which have girths of less than 2m, at least some of which must post-date 1900 (Figure 33). These trees are not scattered evenly across the county. Instead they are clustered in one small area on the southern clays, within a triangle with Attleborough at its apex and Garboldisham and Needham at its base (Figure 34). What is particularly striking is that the youngest *ash* pollards – again roughly defined as those with girths less than 2m – are concentrated in precisely the same restricted area. This is a district characterised by particularly extensive tracts of level, poorly draining Beccles Association soils, but any relationship with environmental factors is clearly indirect. This was the part of the county where small owner-occupiers, as opposed to large and medium-sized landed estates, remained prominent into

FIGURE 33 (left). A diminutive oak pollard, with a girth of just under 2m, at Dickleburgh in south Norfolk.

FIGURE 34. The distribution of young oak pollards (with girths of less than 2m) and young ash pollards (with girths less than 2m) in Norfolk. Soil key: see Figure 2. Dots represent single specimens or small groups of trees.

FIGURE 35. The distribution of young oak pollards (with girths of less than 2m), ash pollards (with girths of less than 2m) and landscape parks shown on the Second Edition Ordnance Survey 6-inch map of *c.*1905. Note how the principal cluster of young traditionally managed trees seems to be associated with one of the principal *lacunae* in the distribution of parks.

the twentieth century, and the concentration of young pollards thus fits neatly into the western half of one of the most significant *lacunae* in the distribution of landscape parks in Norfolk in the early twentieth century (Figure 35) – its failure to fill the eastern part of this space probably the consequence of large-scale hedge removal here in the second half of the twentieth century. This is a particularly telling example of the way in which social, economic and tenurial factors can structure the smallest details of our 'semi-natural' landscape.

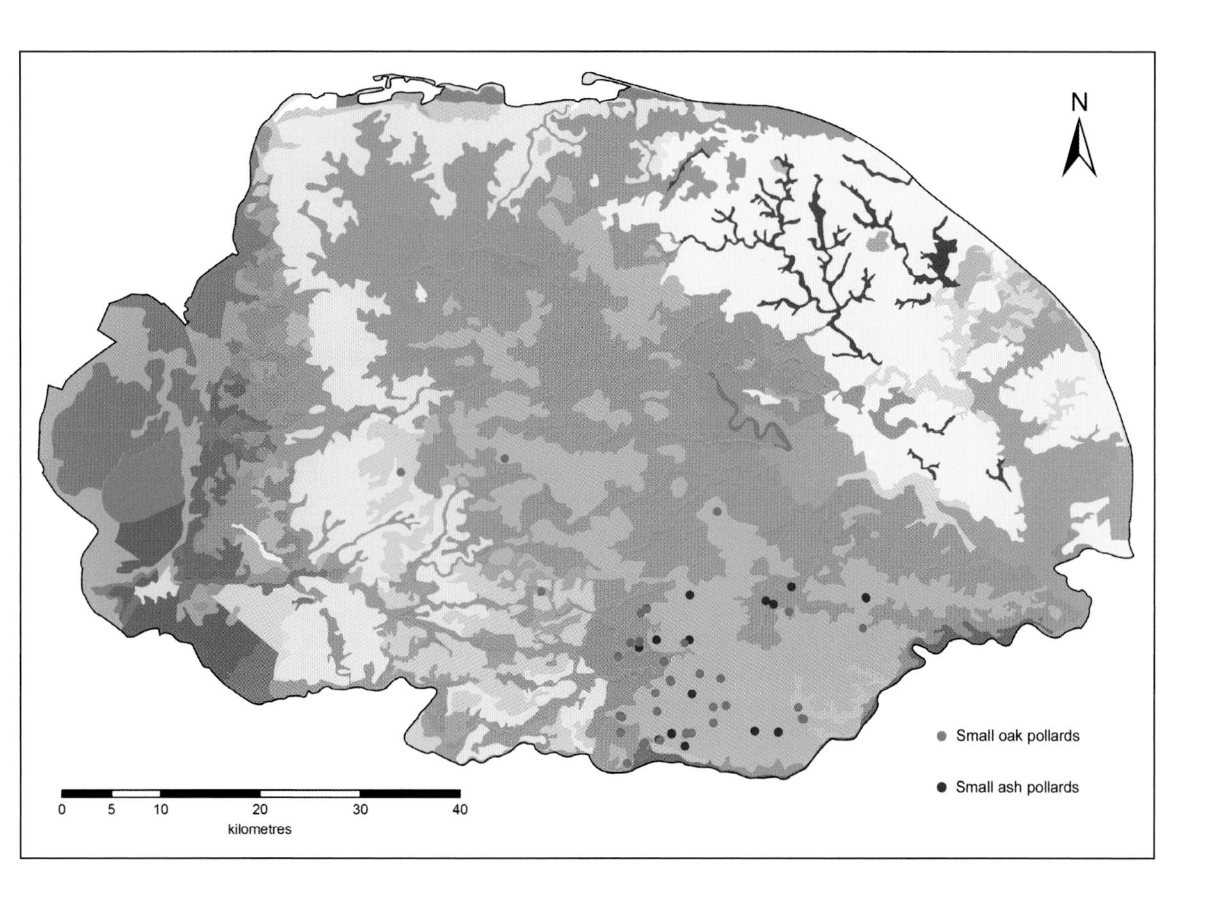
N
Small oak pollards
Small ash pollards
0 5 10 20 30 40
kilometres

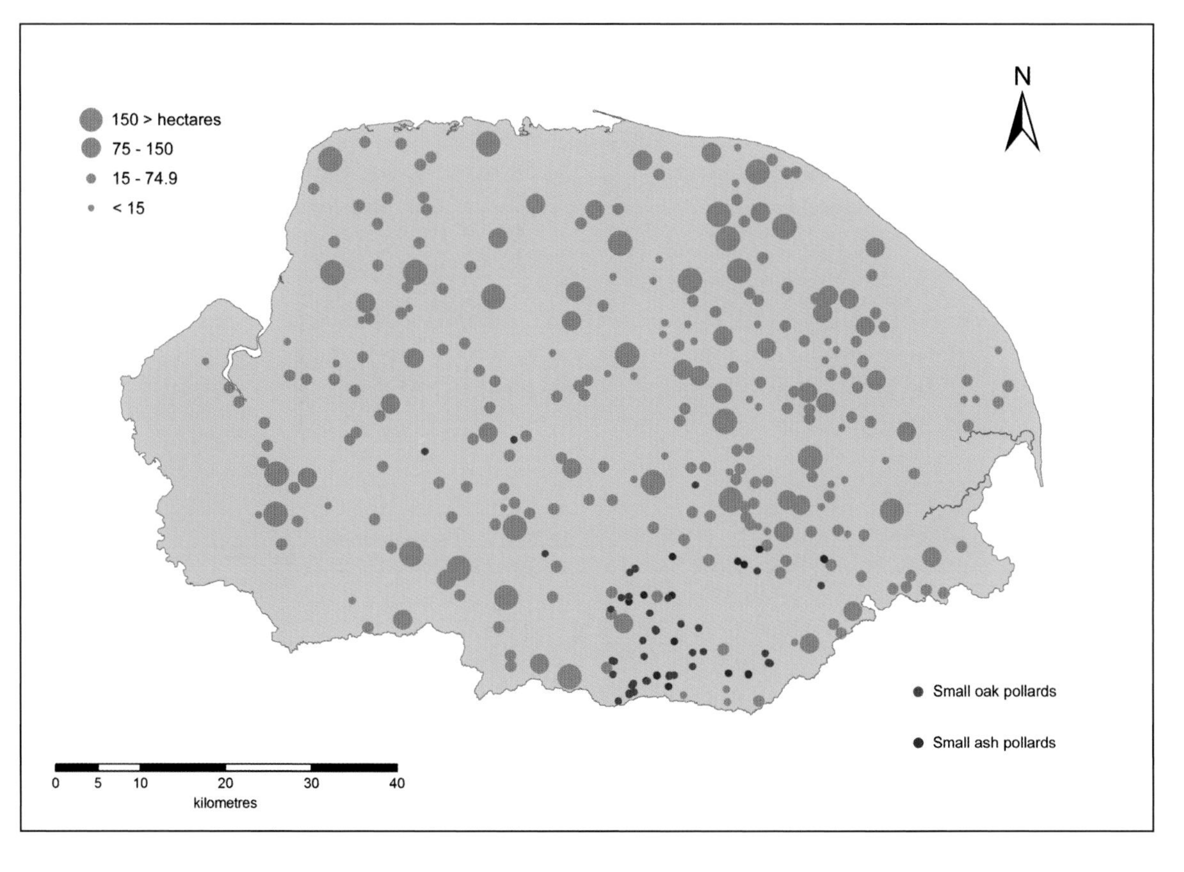
N
150 > hectares
75 - 150
15 - 74.9
< 15
Small oak pollards
Small ash pollards
0 5 10 20 30 40
kilometres

Conclusion

The majority of veteran and traditionally managed trees in Norfolk are found in farmland, and their character poses a number of interesting problems of interpretation. The vast majority are oaks and, on balance, it appears that while the scale of this predominance is partly a consequence of this species' longevity, oak probably was in most districts the most common tree in the early modern landscape. Ash may have come a close second over much of the county but elm was never common and the other kinds of tree sporadically found as pollards or veterans always seem to have been few in number. The fact that the majority of old farmland trees are pollards is, in a similar way, only in part a consequence of differential survival rates; it mainly reflects the fact that most trees in the early modern landscape were managed in this manner, with only a minority allowed to grow as standards. Precisely how the poles from pollarded oaks (and to some extent those cut from other species) were used in the early modern period remains uncertain, although in many cases pollarding probably took place on a relatively long rotation, producing poles substantial enough to be used as small timber or as large fuel logs; while in the case of some of the older trees a short rotation, to produce fodder, may sometimes have been employed.

Perhaps the most important thing we can learn, in historical terms, from an examination of farmland trees is that in spite of what is often suggested pollarding continued on some scale (in spite of the strictures of agricultural 'improvers' and land agents) right through the nineteenth and into the twentieth century. In this and other ways, old trees are in themselves an important source of historical information, telling us about matters scarcely touched upon by documentary sources.

CHAPTER FOUR

Woods and Wood-Pastures

Ancient woodland

Old and veteran trees are rarely found in woods, if by 'trees' we mean full-grown specimens, with a fully developed trunk or bolling. But a small and important minority do exist there and, more importantly, much about Norfolk's silvicultural heritage is difficult to understand without a brief account of the county's woodland history, although this is not our main concern in this book. We need to begin, however, by reiterating the distinction made earlier, between 'traditional' forms of woodland, of the kind which were universal in the medieval period, and plantations. The former were managed as 'coppice-with-standards', the majority of trees and bushes within them being cut down to at or near ground level on a rotation of between seven and fifteen years in order to provide a regular crop of 'poles'. The plants then regenerated vigorously from the stump, or stool, or suckered from the rootstock (Figure 36). Woods like this, in the Middle Ages at least, contained relatively few standard trees,

FIGURE 36. Wayland Wood near Watton is still managed along traditional lines, as coppice-with-standards, by the Norfolk Wildlife Trust.

for if these had been numerous the canopy shade would have suppressed the growth of the underwood beneath. They were normally felled at around eighty to a hundred years of age, when their growth rate began to decline and they were of suitable size to be used in the construction of buildings or – more rarely – ships. Plantations, in contrast, which only began to be established on any scale in the county from the early eighteenth century, consisted entirely of timber trees, without a coppiced understorey, which were planted at the same time, progressively thinned, and then either felled for profit or allowed to grow on, as aesthetic additions to the landscape or as game cover (Rackham 1986a, 64–7; Williamson 1995, 124–40; Daniels 1988).

Ancient woods represent the much-modified remains of the natural vegetation of the county, which probably comprised fairly open and discontinuous woodland rather than dense forest. Whatever its precise character, it had already been extensively altered by human exploitation long before the start of the Middle Ages. By late prehistoric times Norfolk was already a well-settled land, and farms and hamlets were widespread on all soils, including the heaviest clays. During the Roman period settlement intensified still further (Williamson 1993, 42–4; Davison 1990, 15–16). The population evidently declined in the immediate post-Roman period, however, and early Anglo-Saxon settlement sites known from archaeological surveys are much fewer in number and largely absent from the heavier soils. Nevertheless, to judge from the evidence of pollen analysis, there was no significant regeneration of woodland. Although large amounts of woodland doubtless survived right through the Roman period and into the Middle Ages, especially on the heavier clays and higher interfluves, this would already have been altered by grazing pressure and selective extraction, and exploitation intensified once again as population growth resumed in the later Saxon period. By the time of Domesday Book Norfolk was one of the most densely settled regions in England and pressure on woodland must have been intense. Many areas later occupied by coppiced woodland may at this stage have been more open ground, with large numbers of trees (some perhaps already pollarded) but with less in the way of shrubs and underwood, and interspersed with open glades and more extensive areas of grass and bracken.

Woodland contracted steadily through the late Saxon period and by the time of Domesday Book the largest concentrations were to be found in an arc running through the centre of the county from north-east to south-east (Figure 37). This distribution partly corresponds with areas of particularly heavy or acidic soils, but it also follows the high central watershed which divides those Norfolk rivers draining eastwards, into the North Sea at Yarmouth, from those with their outfalls on the north coast or in the Wash. The fact that Domesday Book records woods in terms of the number of pigs that could be grazed there indicates, clearly enough, that most of this comprised wood-pasture rather than coppiced woodland.

These wooded areas continued to dwindle through the twelfth and thirteenth centuries, encroached upon by expanding farmland, which initially took the

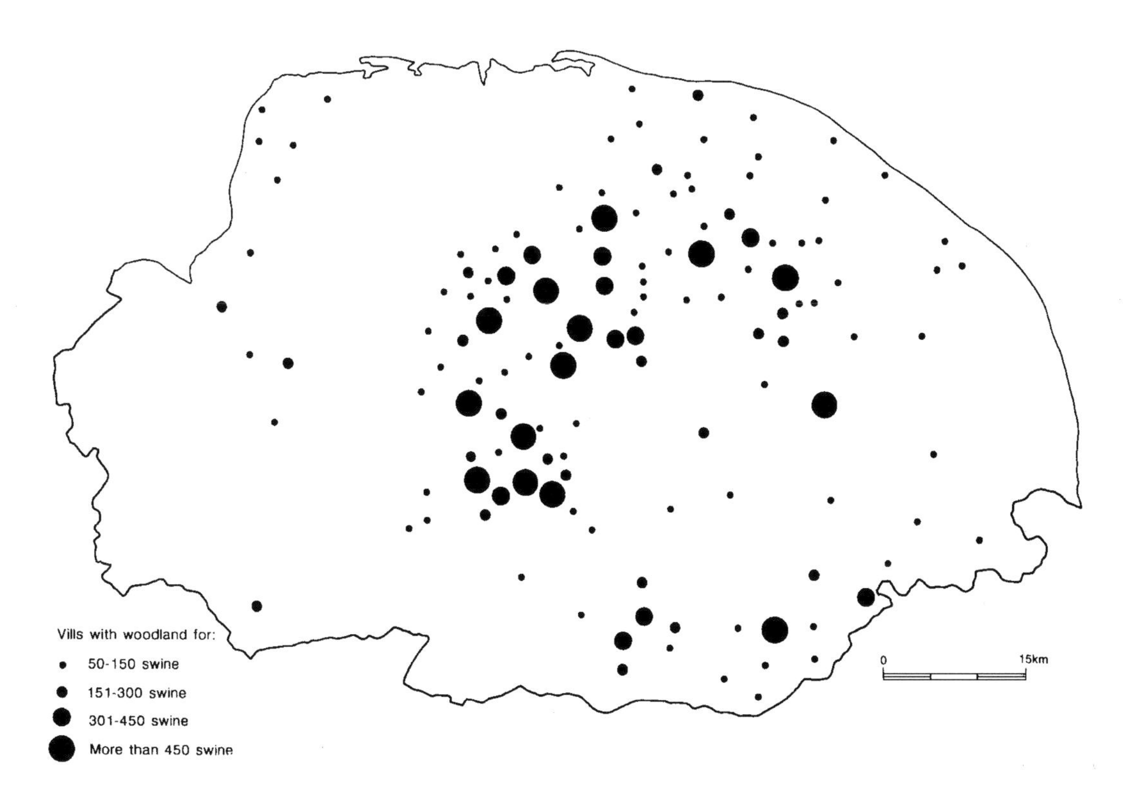

FIGURE 37. The distribution of the main areas of woodland in Norfolk in 1066 according to Domesday Book.

form of additions to the open fields but later usually comprised hedged closes, individually owned and occupied. Those areas that survived unploughed, generally located on the least fertile or most poorly draining soils, normally became areas of common land, to the margins of which farms and cottages were increasingly attracted, forming loose girdles of settlement by the twelfth century. But some areas of the dwindling 'wastes' were enclosed by manorial lords to form deer parks – private wood-pastures used as venison farms and hunting grounds – or private woods. Enclosure not only prevented unregulated felling but also allowed more intensive production of wood through coppicing, for coppice stools could be more closely spaced than pollarded trees. Coppicing was impossible on open land, for livestock would browse off the young regrowth and kill the stools, and so substantial banks, flanked by a ditch and usually topped with a hedge or fence, were constructed to keep them out (Figure 38). A document of 1226 thus describes how the lord of the manor of Bradenham had 'about the wood ... raised one earthwork for the livestock, lest they eat up the younger wood' (Rackham 1986b, 168). Coppiced woods were mainly to be found on the clay soils in the centre, south and east of the county. Here they were often located towards the margins of the main clay masses, probably to limit the distance which wood and timber needed to be hauled along muddy, unsurfaced lanes. Others, a minority, were located on acid sands and gravels, especially in the area to the north of Norwich.

FIGURE 38. Medieval woods were surrounded by substantial woodbanks with a deep outer ditch. Owing to the wood's subsequent expansion this example, at Hockering, now lies well within the area of the wood.

Medieval woodland came in a great range of sizes. As well as the more extensive woods there were numerous groves, springs or 'grovetts', some composed entirely of coppiced underwood, which might cover as little as half an acre. We do not know how many woods existed in the Middle Ages, but we do know that some did not survive until the end of the medieval period. The population, and therefore food prices, continued to rise rapidly until the start of the fourteenth century, and private woodland could easily be grubbed up and turned to farmland at the whim of the owner. Rackham has noted how three of the woods in Forncett mentioned in early medieval documents had disappeared by the fourteenth century (Rackham 1986b, 169). But when the population declined once more following the Black Death and associated disasters of the fourteenth century the number of woods, and certainly the area of many existing woods, increased once again, as some marginal land was abandoned for agriculture and as higher disposable incomes increased the demand for wood and timber. From the sixteenth century, however, population growth resumed and there were further losses, especially during the eighteenth and nineteenth centuries. It has been estimated that around 100 woods, or fragments of woods, of medieval date still survive in Norfolk, although in very varying condition (Rackham 1986b, 170).

There are clear signs that the management of Norfolk's coppiced woods changed over time. In particular, the length of the coppice rotation gradually increased, from around six or seven years in the thirteenth century to around fourteen by the nineteenth. Most of this increase seems to have occurred in the period from *c.*1550 but it has never been fully explained. It may in part have been due to changes in the demand for fuel. The shift from open hearths to chimneys in the course of the sixteenth and early seventeenth centuries increased the demand for larger logs, rather than faggots. But it was also probably because the density of standard trees in woods increased, especially during the eighteenth and nineteenth centuries, from around seven to around twenty per acre (from around seventeen to fifty per hectare), as the increasing availability of coal, and of artefacts manufactured out of iron, reduced the demand for and thus the price of poles relative to that of timber. With more trees, more canopy and more shade, the vigour of the coppice understorey was reduced and the frequency with which it could be cut consequently declined (Barnes 2003, 214–15; 226–8; 294–307). The officials who compiled the documents called the Tithe Files in the 1830s commented on a number of occasions on the poor condition of local coppices, noting at Fulmodeston, for example, that the underwood would have been 'much better if timber was thinned' (TNA: PRO IR 29/5937).

By this time large landowners had been busy for over a century planting new woods, mainly but not exclusively in the form of plantations – stands of trees without a coppiced understorey. By the end of the eighteenth century, to judge from William Faden's county map of 1797, these were widely established, but mainly across the north and west of the county, thus complementing the distribution of ancient woods, which were mainly concentrated in the centre and south-east (Figure 39). Such woods were planted in part for profit, but also to beautify the landscape and to provide shelter for game birds, especially pheasants. William Kent, writing in 1796, observed that while 'gentlemen of fortune' in Norfolk had carried out much tree-planting 'in their parks and grounds', the planting of 'pits, angles, and great screens upon the distant parts of their estates, which I conceive to be the greatest object of improvement, has been but little attended to' (Kent 1796, 87). Kent exaggerated slightly: although many of the new woods were indeed close to parks and mansions, nearly 40 per cent (to judge from Faden's map) were more distantly placed. The composition of these new areas of woodland was often very different from that of traditional woods. Although oak was widely planted, beech was also a major timber tree, as were elm, sweet chestnut and sycamore, while a range of conifers were employed either as 'nurses' (interplanted with the hardwoods to give them shelter, and progressively thinned as the plantation matured) or to form plantations in their own right. By the second half of the eighteenth century a wide range of species was available from commercial nurseries in Norfolk, some intended as single specimens for pleasure grounds but mostly destined for plantations. In February 1776 a sale was advertised in the *Norfolk Chronicle* at 'Mr Hardy's old nursery ground at Catton' of various trees, 'very cheap'. They included 'fine four years

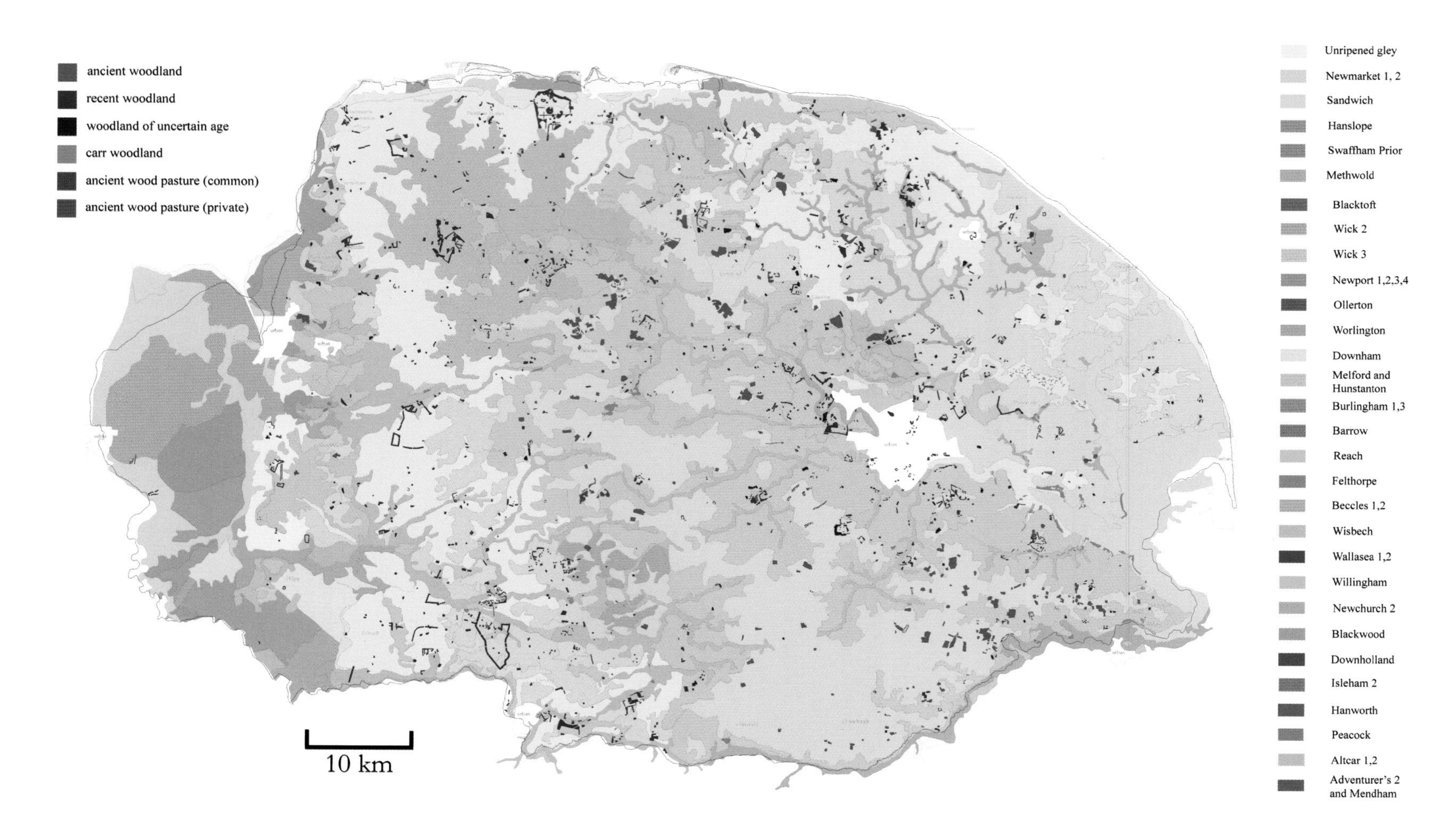

FIGURE 39. The distribution of different kinds of woodland in the 1790s, based on William Faden's county map of 1797 (the map does not itself distinguish between different kinds of woodland: the classification is based on an examination of historical evidence and locational and botanical characteristics).

old Scotch firs (being the best Age for planting on Heaths or Grounds)'; Lord Weymouth pines; Cluster pines; pinasters; spruce; Balm of Gilead fir; Carolina pitch pine; 'Spanish' (i.e. sweet) chestnut; ash; sycamore; lime; 'occidental planes'; arbeels (black poplar), 'English and foreign poplars, Hertfordshire, Dutch and narrow-leaved English elms', as well as oak and beech.

In spite of the popularity of the new forms of forestry, most ancient woods were still being managed along traditional lines right through the nineteenth century. In 1851 Henry Wood, agent for the Merton estate in Norfolk, described in some detail the exploitation of a seventy-acre wood (almost certainly Wayland near Watton) and listed the varied uses which were made of the coppice poles, which were principally of ash and hazel. Everything was either sold or utilised on the estate. Larger poles were employed to make hurdles, garden fencing or bins for storing hay or straw on the estate's home farm; splints, about 2m in length, were used for building work; smaller material went for thatching broaches and sways, or for pea sticks; the smaller brush faggots were 'sold to bakers and cottagers for oven wood'; while the off-cuts were sold for cottage firing (NRO WLS XVIII/7/1). Moreover, a small number of entirely new areas of coppiced woodland were created in the course of the eighteenth and nineteenth centuries; these can sometimes be hard to distinguish from genuinely ancient woods, apart from the fact that they generally have small woodbanks (or no woodbank at all). Lopham Grove, in the extreme north of North Lopham parish, contains a mixed coppice understorey dominated by hazel but with areas of hornbeam on the damper soils and scattered examples of maple. There are oak standards and a substantial bank, coinciding with the parish boundary, runs along two sides: pollarded hornbeams grow on the more diminutive southern boundary. In spite of appearances, the map evidence leaves no doubt that the wood was created, at the expense of pasture fields, only at some point after 1754 (Arundel archives). Definitions can be blurred and confusion arise in other ways. Not surprisingly, given their common ancestry, a large number of Norfolk woods abutted directly upon commons, which were usually by now only sparsely treed. In the late eighteenth and early nineteenth centuries most of these were enclosed and turned into private property, usually through parliamentary acts. The majority were then 'improved': that is, they were converted to better-quality pasture or (more usually) arable land, but some were planted up with trees by wealthy landowners. Further additions to existing woods were often made in the late nineteenth century, when the severe agricultural depression encouraged the abandonment of farming on the heaviest clays and the poorest of the light soils in Norfolk. As a consequence of these developments a central 'core' of ancient woodland is often surrounded by areas of eighteenth- or nineteenth-century expansion, usually comprising plantations but sometimes featuring new areas of coppice. Edgefield Little Wood in the north of the county is something of a misnomer, as it today covers around twenty-nine hectares. Closer examination reveals that the original medieval wood, comprising oak coppice within a substantial perimetre bank, was expanded in the course of the nineteenth century across former heathland to the east, the new additions partly comprising sweet chestnut coppice.

It was only in the course of the twentieth century, and especially during the inter-war years, that traditional forms of woodland management generally came to an end (Wade Martins and Williamson 2008, 130–32). This was in part because they had become unfashionable among landowners and in part because, as large estates were broken up, many woods fell into the hands of local farmers who, as Butcher said of their fellows in Suffolk, 'know little and care less for forestry' (Butcher 1941, 361). But it was also a consequence of the continuing contraction in demand for coppice poles, as alternative materials for fencing, tools and the rest became available owing to improvements in transport. Coal was available at relatively low cost and household grates and ranges were adapted accordingly. In the middle decades of the twentieth century a number of ancient woods were grubbed out altogether and turned over to farmland, or were replanted with commercial conifers (usually with limited success). Only a few continued to be managed along traditional lines. Since the 1960s, however, there has been a significant resumption of coppicing, now carried out by landowners or conservation groups keen to maintain the diversity of flora and fauna produced by this kind of management.

Ancient woods are important archaeologically. Many contain earthworks, some relating to the management of the wood itself, some to activities which took place on the site it occupies before it came to be managed as coppice (Rackham 1976; Rotherham and Ardron 2006; Morris 2003; Bannister 1996). Hockering Wood, for example (Figure 40), contains a moated site, a large trapezoid enclosure of uncertain date, a feature which may be the remains of a prehistoric linear earthwork and complex banks and ditches which represent divisions of use or ownership, as well as phases of expansion and contraction. There are earthworks of bomb stores and Nissen huts, created in the Second World War when the wood was a forward ammunition dump for local airfields. In general, such remains support or amplify the story we have just briefly sketched out earlier. Several woods thus contain earthworks which seem to relate to the management of wood-pastures in the early medieval period, before the wood was enclosed and separated off from the surrounding tracts of wooded 'waste'. In particular, in a numbers of places earthworks of substantial manorial settlements and enclosures are found, either within ancient woods or in places known to have been cleared of such woodland in the nineteenth century. Examples include the extensive complex of demesne enclosures at Horningtoft (Cushion and Davison 2003, 111); the motte-and-bailey castle, and associated trapezoid enclosure, at Denton (Cushion and Davison 2003, 168); the moat, associated with a similar enclosure, in Hockering; and the two sites – one a simple moat, one an irregular and partly moated enclosure – within Wood Rising Wood (NHER 25241). The castle at Denton is particularly interesting in this respect. It was raised, probably during the 'anarchy' of 1135–54, on a remote property of the d'Albini family, not far from the great *caput* of their rivals, the Bigods, at Bungay. It stood some way from any

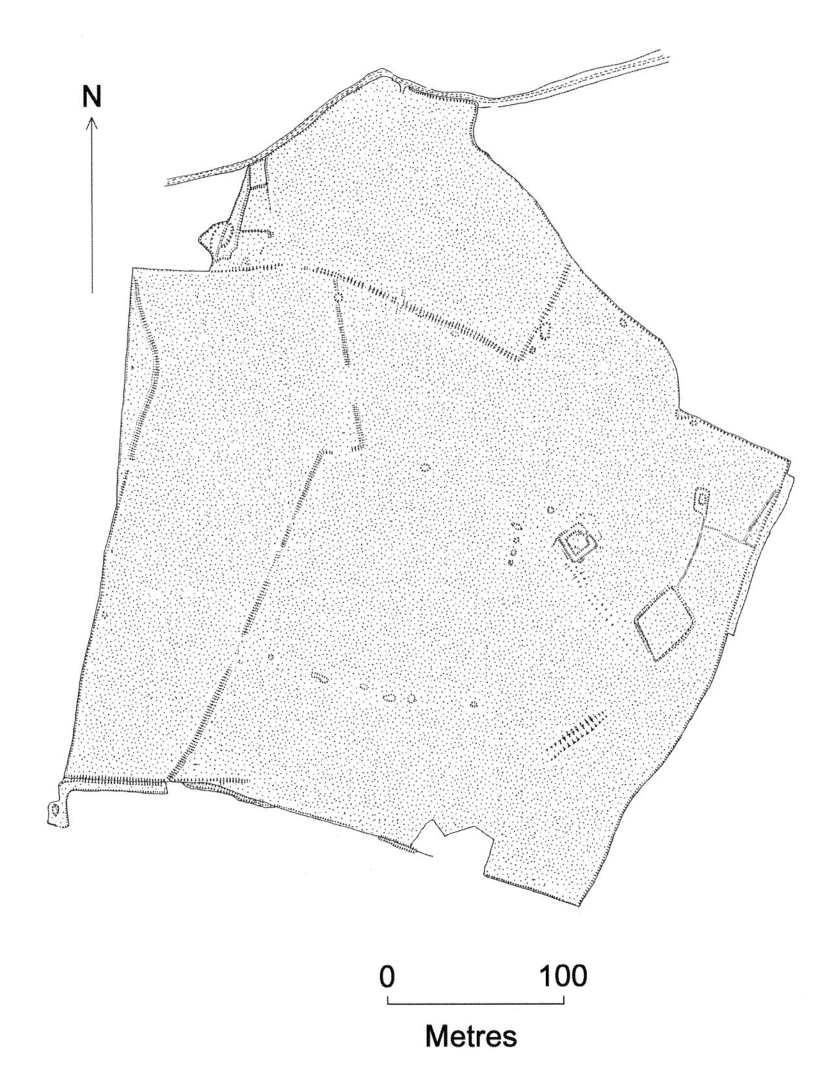

FIGURE 40. Hockering Wood contains a remarkable range of earthworks. These include numerous banks marking where the wood has expanded, or has been divided between different owners or management regimes; a medieval moated site and associated ponds in the centre/east; a trapezoid enclosure of uncertain date and purpose to the south-east of this; and, to the south-west, a substantial feature which may be a hollow-way or former road, but which could be the remains of a linear earthwork of prehistoric date. (The earthworks relating to the use of the wood as an ammunition store in World War II have been ommitted.)

significant settlement, but was clearly built to defend something. While wood and timber are not obvious targets for enemy raids, cattle are. Most of these sites were probably associated with the protection and management of demesne herds within extensive areas of unenclosed woodland grazing.

More numerous are earthworks which date to the time when the wood was enclosed and coppiced. Medieval woods were, as already noted, surrounded by substantial banks originally topped by a fence or hedge and accompanied by deep outer ditches. These earthworks are generally much more massive than those which were used to keep livestock out of later woods, and they are thus hard to explain in purely practical terms. They may, in part, have had a symbolic significance, asserting lordly possession of what had formerly been a part of the common wastes. Many woods also have internal boundaries, marked by similar earthworks, which represent either where ownership of a

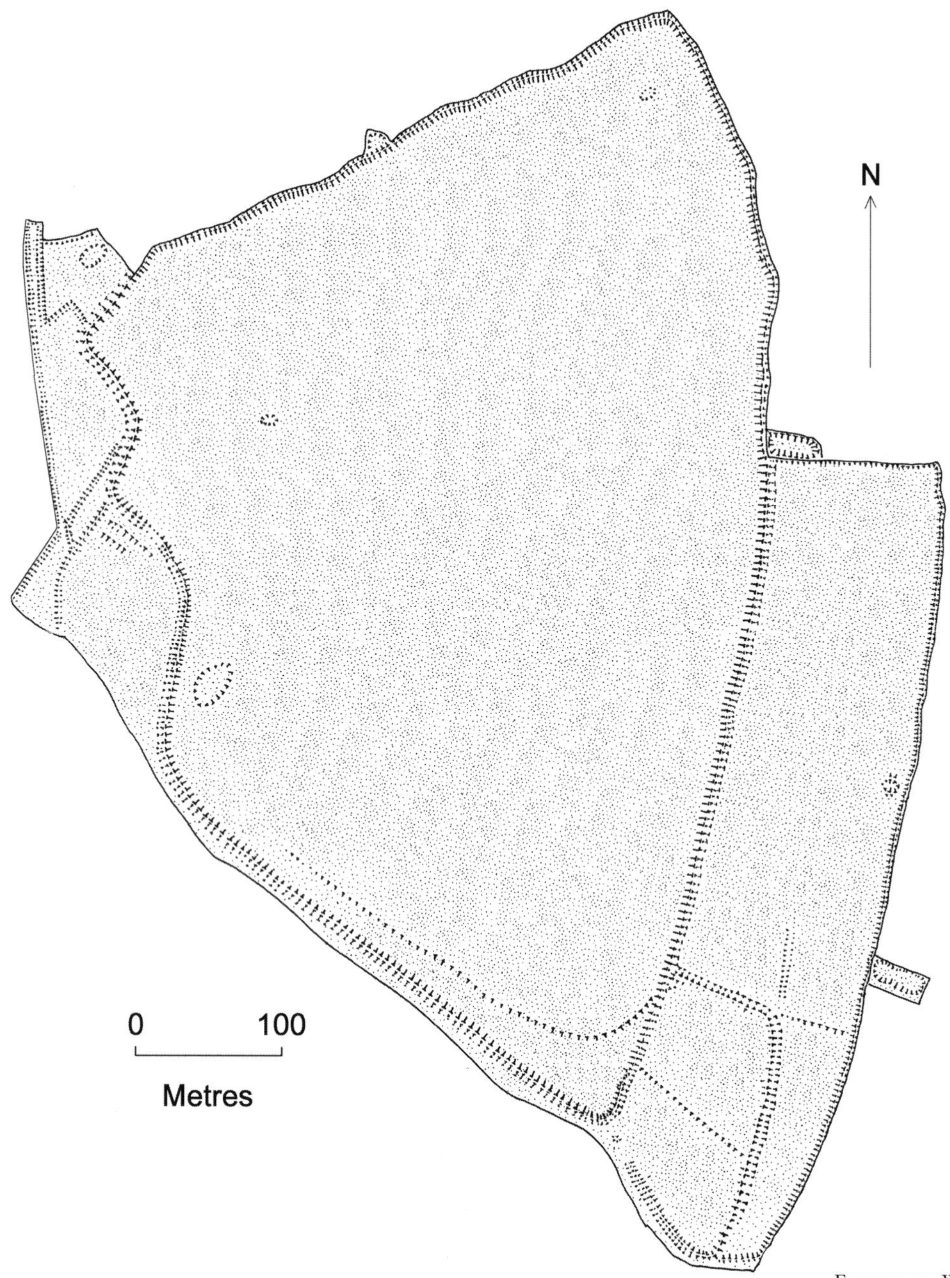

FIGURE 41. Wayland Wood, Watton. The majority of earthworks within the wood represent woodbanks marking where the area of the wood has grown, in stages, in late medieval and post-medieval times.

wood became subdivided or where portions were managed in different ways – one by coppicing, one perhaps still exploited as wood-pasture. In other cases, internal banks mark where woods have been expanded at some point in the past. The configuration of woodbanks leaves no doubt that Wayland Wood near Watton, for example, was extended to the south-east at some point in the past, probably in the fourteenth or fifteenth centuries (Figure 41), and it

FIGURE 42. Massive stools of hornbeam and ash growing in the eastern extension to Wayland Wood.

is noteworthy that the vegetation of the 'extension' is still distinctly different from that of the 'original' wood. The latter is characterised by an understorey of ash, maple, hazel and bird cherry (an unusual species in Norfolk, at the southern edge of its range). The extension, in contrast, while also featuring ash and some hazel, boasts numerous massive stools of hornbeam (Figure 42), here some way to the west of the main concentration of this species in woodland, in the south-east of the county. Additions to woods often took the form of narrow strips, creating two parallel woodbanks, as at Hockering (on the western side), Sporle and Shotesham Little Wood. In this last case the vegetation within the extension, composed entirely of coppiced hazel with oak standards, is still markedly different from that in the main wood, which contains hazel, birch and ash, but is dominated by hornbeam.

The vegetation of ancient woods in Norfolk

This brings us to our main concern – the character of the trees found in woods, and how the development of Norfolk's woodland fits more generally into the county's wider arboricultural history. Silviculturalists and foresters often use the term 'stand type' to discuss the particular combinations of trees and shrubs found in woods. In most ancient woods in Norfolk a combination of stand types can be found. Traditionally, the full-grown standard trees were mainly oaks (Rackham 1986b; Barnes 2003). The coppice understorey, in contrast, varies greatly in composition, although across a wide area of the county is dominated by some mixture of ash, hazel and maple. Hazel was particularly useful for hurdles and fencing, and for the wattle-and daub used in the walls of timber-framed buildings; while ash, as we have seen, grows rapidly and makes excellent firewood, as well as having a wide range of other practical uses. In particular parts of Norfolk, however, other species are prominent (Rackham 1986b; Barnes 2003). In the south-east of the county, and extending into north-east Suffolk, most woods are dominated by hornbeam. Small-leafed lime (*Tilia cordata*) occurs in a number of places in the centre and centre-north of the county, most notably in Hockering Wood and to a lesser extent in Swanton Novers Great Wood. Bird cherry is a major component of Wayland Wood, and of some other woods on the edge of Breckland, while coppiced oak occurs in a few woods on the poorer soils in the north of the county, including Edgefield Little Wood and (again) parts of Swanton Novers Great Wood.

As yet we have no very clear idea of the reasons for these variations, which are not always very closely related to environmental factors. Norfolk's lime woods are thus found on clay soils of the Beccles and Burlingham Associations, but only within a very restricted section of the area which they occupy. Other woods, lying no more than 5km away and on the same soils, contain no lime. The far more numerous hornbeam woods are similarly mainly restricted to Beccles and Burlingham soils, but mainly in the south-east of the county: similar soils extend in a broad band through the centre of the county and almost to the north coast, but much less hornbeam coppice is to be found here (and many of the exceptions, as at Wayland, appear to be in relatively late additions to existing woods). This issue is made more complex by the fact that the present predominance of hornbeam in such woods is almost certainly, in part at least, a consequence of quite recent changes. Before the late nineteenth century the vegetation of these woods was probably more mixed in character, to judge from those portions (as at Sexton's Wood in Denton) which are still coppiced, or in which management has declined comparatively recently. The gradual reduction in management in the course of the twentieth century and the consequent growth of the hornbeam to canopy level has probably served to shade out other understorey shrubs such as hazel or maple. But such relatively recent changes, important and often neglected though they undoubtedly are by students of woodland ecology, have really only served to accentuate existing patterns of vegetational variation, and cannot explain them. Presumably the traditional composition of these woods reflects a complex interplay of environmental factors – soil type, drainage – and management history, with landowners or their agents deliberately encouraging (or discouraging) particular elements of the natural vegetation over the centuries. Lime was thus presumably a significant element of the natural vegetation around Hockering Wood but was then actively encouraged, to the point of localised dominance. The string of ponds found within the wood, associated with the moated enclosure, was perhaps used for retting bast, the fibres from lime bark which were widely used in the early Middle Ages to make rope. The wood may have functioned, in effect, as a rope factory (although once again we need to ask whether the present overwhelming dominance of lime within the wood is, in fact, the result of relatively recent developments since the mid-nineteenth century – the neglect of 'traditional' management, and the deliberate encouragement of lime as a single species).

Woods, in other words, are less 'natural' than we often assume. The pollen evidence suggests, as we have seen, that the trees which dominated the natural vegetation of Norfolk differed significantly from those found in the county's surviving ancient woods. In particular, small-leafed lime (*Tilia cordata*), which is now confined to a small number of woods, was then apparently a fairly common tree, while hornbeam was, to judge from the available evidence, almost unknown. The character of the natural 'wildwood' had probably been significantly altered by human exploitation long before many of its surviving fragments were enclosed and transformed into managed coppice-with-standards woodland. Lime, for example, does not respond well to grazing pressure, and was replaced by other species, particularly hornbeam – the latter increasing its importance as early as the Roman period, if the pollen sequence from Diss Mere has been correctly dated (Peglar *et al.* 1989). Subsequent decisions by woodland managers over which species on a site should be encouraged over others presumably led to further changes; so too did a wide variety of 'natural' processes, such as the spread of elm, by suckering, from adjacent hedges or standard trees, as apparently in Hedenham Wood. It is not surprising, then, that the stand types found in Norfolk's 'natural' woods differ in significant ways from the range of plants found in the unmodified, post-glacial 'wildwood', whatever precise character this had taken.

As already emphasised, few really old trees are found within ancient woods, in the sense of full-grown standard specimens. The oak standards present today are usually larger than they would have been in intensively managed medieval woods, but are still small compared with the kinds of veteran trees, especially large old pollards, which can often be found on the surrounding farmland. Trees in woods were usually felled at around eighty to a hundred years, so the age of the oldest woodland trees is normally equivalent to this plus the amount of time that has lapsed since the wood was last systematically managed on an economic basis – often the end of the nineteenth century, but sometimes earlier. Few standard trees in woods are thus more than two or three centuries old, and most are considerably younger than this. But some coppice stools can be very ancient. Although the individual stools of some species do not live very long, those of others – especially ash, and to a lesser extent hornbeam – can survive for centuries, forming great spreading rings of vegetation. In addition,

FIGURE 43 (left). At Edgefield Little Wood the perimeter woodbank is still topped by an outgrown oak hedge of medieval date.

FIGURE 44. A pollarded hornbeam on the northern boundary of East Wood, Denton, south Norfolk.

the substantial woodbanks surrounding medieval woods were often surmounted with a hedge and while in most cases this has long since disappeared, shaded out by the trees, there are some notable exceptions. In some south Norfolk woods the hedge was composed of hornbeam which, being resistant to shade, thus survives in recognisable form, as at Shotesham Little Wood. More impressive is the medieval oak hedge which tops the perimeter bank all along the southern and western edges of Edgefield Little Wood in the north of the county, now grown into a series of massive, disconnected stools (Figure 43). At both places the hedge is composed of the same shrub which dominates the coppiced understorey within the wood, and this may well have been normal practice (what may be the remains of a lime hedge survive along part of the Hockering Wood's boundary). Most woodbanks lack clear evidence for relict hedges, especially in the central parts of the county, and this is perhaps because hazel, maple and ash survive prolonged shading less well than oak or hornbeam. Others were doubtless surmounted by fences.

In most woods the only true 'trees' of any real antiquity are the pollards which were often planted on the woodbank surrounding the wood. These are often oak, but alternatives again tend to replicate the character of the vegetation within the wood: notable examples include the striking hornbeams on the banks of East Wood, Denton (Figure 44), and the lime pollards on the southern edge of Hockering Wood (Figure 45). Some of these may be medieval trees, but most are post-medieval, and there is no doubt that new pollards continued to be established on wood boundaries even in the eighteenth century – by which

FIGURE 45. A pollarded lime tree on the southern boundary of Hockering Wood in central Norfolk.

time these were often banks and ditches no larger than those associated with normal field boundaries. Wayland Wood was thus expanded to the west some time after 1724, to judge from a map of that date (NRO WLS XVII/9 410X6), and an oak pollard which stands on the diminutive boundary bank has a girth of 3.5m, consistent, in such a location, with an eighteenth-century planting date.

One last, rather specialised, type of woodland needs to be briefly mentioned: wet woodland, or carr, an important but often neglected component of Norfolk's historic landscape, which occurs on the floors of many river and stream valleys in the county. It comes in a variety of forms, depending in large part on the extent to which the site is waterlogged, but is always dominated by alder, with varying amounts of willow and (in the drier locations) birch, ash and oak. Woodland of this kind features on numerous seventeenth-century maps, such as Thomas Waterman's survey of Raynham, made in 1617 (NRO BL 33). As already noted, alder was an important wood: if kept underwater it is remarkably resistant to rot, making it ideal for piling and for the construction of jetties. Alder poles are also straight and firm, making it a useful material for scaffolding; while when burnt in the correct circumstances the wood makes premium charcoal, ideal for the manufacture of gunpowder (in 1612 Old Carr Wood in Gressenhall was leased for two years by the Le Strange family of Hunstanton: the rent included four loads of charcoal, delivered to Hunstanton during the term of the lease (NRO MR 211 241X6)). Although most woodland of this type was coppiced, some of the drier carrs were managed as wood-pastures and grazed in the drier summer months. A number of the Gressenhall 'plantings' depicted on Waterman's map of 1624, discussed below, thus seem to have comprised wet woodland which was grazed as well as being cut for wood (NRO Hayes and Storr 72). Carr woodland is relatively easy to identify on Faden's map of Norfolk, surveyed in the mid-1790s, partly because he often labels it as 'carr' but also because of its location. Around 1426 hectares of the woodland shown on the map was of this type, just under 11 per cent of all the woodland depicted in the county (in addition, in some places the word 'carr' is shown in association with individual tree symbols, scattered across an area of fen, as in parts of the Bure valley, again indicating a form of wet wood-pasture). Yet while carr is evidently an ancient feature of the landscape, and many of the areas shown on Faden's map were probably very ancient at the time it was surveyed, there is some evidence that much was recent, and that in general this kind of woodland is a comparatively transient feature of the landscape. By comparing successive maps it is often possible to see how particular areas of wet woodland develop at the expense of fen or meadow, last for a few generations, and then revert to grass and reeds once more – for reasons which remain uncertain.

Wood-pasture commons

Those areas of common wood-pasture which survived the encroachments of cultivated land and enclosure as coppiced woodland or deer parks into the Middle Ages were, according to many writers, inherently unstable environments. Their trees were vulnerable to damage from stock, through the stripping of bark for example, or the compaction of the ground above their root systems. More importantly, once trees died, were blown down or were felled it was difficult to establish replacements in the face of sustained grazing. According to conventional wisdom, fencing off portions of land to protect new trees (or indeed for any other purpose) was difficult, if not impossible, as it conflicted with the rights of commoners to freely access and exploit the common (Rackham 1986a, 121–2). But careful examination of both the documentary evidence and the location of ancient trees surviving in the modern landscape calls such assertions into question. It is now clear that wood-pastures *were* a common feature of the East Anglian landscape right through the medieval period, and – more surprisingly – that in some places they continued to exist, and to be actively managed, into the eighteenth century and beyond.

This survival was partly due to the fact that, as Patsy Dallas has shown, management systems often existed which allowed or encouraged the replacement of lost trees on common land (Dallas 2010a). That is, many local communities developed customs which permitted commoners to plant new trees on commons and to protect them during the early stages of growth. Such arrangements are well documented in the south and east of the county, where commons were usually ringed by farms and villages. During a legal dispute in the late sixteenth century concerning the commons at Pulham, for example, it was stated that 'The tenantes of the said manor have used to make benefitt of the trees growing upon the common near their houses which were planted by themselves and their predecessors' (NRO NAS II/17). A manorial survey of 1579 relating to the parish of Gressenhall similarly records that when tenants were admitted to holdings they received one or more 'planting' (NRO MR61 241X1). A detailed map of 1624 shows that 'plantings' were areas of scattered trees growing on the various commons of the parish, in some cases so dense that the areas in question were effectively wooded. They are associated by name with the various owners close to whose homes they were located (NRO Hayes and Storr 72; see Figure 52). Such customs certainly continued in many places into the eighteenth century. Francis Blomefield, in his *Topographic History of the County of Norfolk* of 1739, described how the tenants of his home parish of Fersfield had 'Liberty to cut down timber on their copyholds, without licence and also to plant and cut down all manner of wood and timber on all the commons and wastes against their own lands, by the name of an outrun' (Blomefield 1805, vol. 1, 739, 95). Once again the trees established by the commoners were close to their 'own lands'. Blomefield describes similar customs in other south Norfolk parishes, including Kenninghall, Diss and Garboldisham (Blomefield 1805, vol. 1, 220, 263).

A few traces of these wooded commons still survive in the modern landscape, as at Old Buckenham and Fritton. The latter common, a large example ringed (in customary fashion) by a girdle of ancient farms and cottages, boasts no fewer than twenty pollarded trees, many hidden away amidst the younger scrub and woodland which has regenerated in relatively recent times due to a decline in regular grazing. The pollards are concentrated towards the edges of the common for the most part, near to where the houses stand. Most are oaks with girths in the range 2.3–5.4m, but there are six examples of ash (2.3–3.4m) (Figures 46

FIGURE 46. Pollarded oak trees surviving among later scrub and secondary woodland on the edge of Fritton Common, on the claylands of south Norfolk.

Figure 47. A fallen ash pollard on Fritton Common; its comparatively small size suggests that new pollards continued to be established here well into the nineteenth century.

and 47) and, somewhat surprisingly, a sycamore (4.1m). Most of these are not very old trees. At least half, to judge from their size, must have been planted and first cropped in the nineteenth century, and many were probably cut regularly into the twentieth. It is noteworthy that the trees on Old Buckenham common are similar in character: the pollards are again all fairly young, ash is well represented (here in fact forming the majority of trees), while the unusual sycamore at Fritton here has its parallel in a pollarded sweet chestnut. The trees on the green at Saxlingham Nethergate, including pollarded ash, black poplar and sycamore, all of no great antiquity, also bear comparison.

Wooded heaths

Rather different kinds of wood-pasture commons could be found on the dry, acid soils in the north and west of the county. According to much conventional wisdom these areas had generally been stripped of their trees at a very early date, before the Middle Ages. Sandy soils, although comparatively infertile, were easy to cultivate and thus attractive to prehistoric farmers equipped with only primitive implements. Trees here were comprehensively cleared to make way for cultivated land, and any woodland that remained degenerated at an early date, under grazing pressure, to open heath (Figure 48) (Dimbleby 1962). The acid character of these soils, coupled with the continued pressure from grazing animals, then favoured the development of a characteristic undershrub vegetation, dominated by heather (called ling in East Anglia) (*Calluna vulgaris*), bell heather (*Erica cinera*), gorse (furze in East Anglia) (*Ulex europaeus*) and broom (*Sarothamnus scoparius*). This is turn encouraged the development of a particular type of soil, the podzol, in which grey upper levels, leached of humus and iron, overlie hard layers of humus pan and iron pan, where these have been redeposited (Rackham 1986a, 286–91; Dimbleby 1962; Parry 2003). Characteristic grasses also thrive in such environments, including sheep's fescue (*Festuca ovina*), wavy hair grass (*Deschampsia flexuosa*) and common bent (*Agrostis tenuis*), while some areas become dominated by bracken. In some parts of Norfolk chalk heath, in which grasses like sheep's fescue are dominant and the various undershrubs constitute subsidiary elements, could also be found, normally where thin and intermittent sandy deposits overlie chalk (Grubb *et al.* 1969). Elsewhere, and especially in the central and northern parts of the county, heaths occupied areas of acid gravel, as well as sand.

Whatever its precise character, heathland played an important role in the medieval and post-medieval economy. Not only were the heaths grazed, by sheep and in some cases by rabbits, but, in addition, they were regularly cut for gorse (used for fencing and fuel) and for bracken and heather (used for thatch and cattle bedding). In many areas of Norfolk, moreover, some of the heaths were ploughed – either sporadically, or on a long rotation, as outfield 'brecks'. Not surprisingly, such intensive exploitation kept the heaths open and treeless. Heaths do not therefore constitute a stable 'climax' vegetation but are a

FIGURE 48. Typical Norfolk heathland at Salthouse on the north Norfolk coast.

consequence of particular forms of management, and when grazing and cutting are reduced, or cease altogether, they begin to change rapidly and become invaded by hawthorn, sloe and birch, and eventually by oak. It is difficult for trees and bushes to colonise dense stands of heather. But the older, more degenerate stands which develop once grazing is reduced provide more open ground. Once the trees become established they shade out the heather, leading eventually to the development of the *Quercus–Betula–Deschampsia* woodland which is the natural climax vegetation on these poor soils (Rodwell 1991, 377).

Many Norfolk heaths were reclaimed in the course of the Middle Ages and converted to permanent arable. But vast tracts survived into the eighteenth century, mainly in Breckland and on the Greensand ridge to the north of King's Lynn, but also in the 'Good Sands' distinct of northwest Norfolk, and on the intermittent strip of acid, gravelly soils running northwards from Norwich to the sea. Scattered examples could also be found on the clays in the centre of the county, associated with localised spreads of glacial sand and gravel. Then, as cereal prices rose and agricultural techniques were improved, systematic attempts at reclamation were begun, reaching a peak during the high price years of the Napoleonic Wars (Gregory 2005). Where deposits of sand were thin over more calcareous formations heaths were converted to arable wholesale, as across much of north-west Norfolk. Elsewhere, in Breckland and in the area north of Norwich, reclamation was less successful. Here, much heathland was never converted to productive farmland, or else reclamation lasted only for as long as grain prices remained high, the land then reverting to rough grazing once again – sometimes immediately after the Napoleonic Wars ended, sometimes as prices fell back more dramatically with the onset of the great agricultural depression from the late 1870s (Wade Martins and Williamson 1999, 34–46). By this time, heathland had lost its former economic value. In the inter-war years large areas were planted up by the Forestry Commission. Much of what remained began to be invaded by scrub and in places has now reverted to secondary woodland in the manner described.

We have presented here, albeit in rather bald and simplified form, the generally agreed history of heathland in England in general, and Norfolk in particular. Many heaths did, indeed, become completely open landscapes in the Bronze or Iron Ages, continuing as such into modern times. Accumulating evidence, however, suggests that many in fact carried areas of managed wood-pasture into the Middle Ages, and sometimes much later. It has long been appreciated that at least one Norfolk heath, Mousehold near Norwich, was woodland in medieval times (Rackham 1986a, 299–302). Domesday Book implies that there was a substantial wood here and the element 'hold' in the place-name derives from *holt*, an Old English term for 'wood'. In the thirteenth century the agent of the bishop of Norwich complained that it was proving difficult to restrain the tenant's use of the wood – which was common land – and the trees were disappearing. By the end of the century, documents refer to Mousehold *Heath*.

It now seems that Mousehold was by no means unique. Several early documents describe wood-pastures in parishes which entirely comprise sandy and gravelly soils and which, by the later eighteenth century, consisted solely of arable fields and heaths. In 1482, for example, Hugh Donne leased the lands owned by Mount Joye Priory in Haveringland and Felthorpe 'except all the woods of the place'. The document includes the clause: 'the said Hugh shall have from the Prior 200 good faggots yearly and sufficient thorns and underwood for fencing the closes ... and all the old wood, boughs, sticks and windfall wood that fortune to fall within the said lands except great trees' (NRO NRS 21788 361X2). In the late sixteenth century the inhabitants of nearby Marsham accused James Brampton of erecting a foldcourse 'where he ought not' and asserted that he had 'felleth downe woode growinge uppon the common contrarye to the custome of the mannor' (Smith *et al.* 1982, 242–3). A number of early maps also show partly wooded heaths, including a survey of New Buckenham made in 1597 (NRO MC 22/11); a map of Haveringland of 1600 (NRO MS4521); an undated sixteenth-century map of Castle Rising (NRO BL 71); and a survey of Appleton of 1596 (NRO BRA 2524/6). Indeed, close examination indicates that a number of residual fragments of wood-pastures are actually shown in the form of scattered tree symbols within areas specifically described as 'heath' on William Faden's 1797 map of Norfolk (as on Walsham Heath, Hevingham Heath, Necton Heath, Stock Heath, Edgefield Heath and Cawston Heath). But more important than the evidence of documents is that provided by the actual remains of some of these ancient heathland wood-pastures still surviving in the modern landscape.

On the Bayfield estate in north Norfolk, for example, more than seventy ancient pollarded oaks – some pedunculate *Quercus robur*, some sessile *Q. petraea* – survive on dry, leached soils of the Newport 4 and Barrow Associations. Some are found to the west of the river Glaven, rather more on steep slopes to the east. The youngest of these trees may be less than 300 years in age – indeed, the majority have girths of less than 5m – but the largest, the so-called 'Bayfield Oak', has a girth of 9.2m and may be around 700 years old (Figure 49). Most are found within estate woodland which was already in place by the late eighteenth century but some are in areas which were still denoted as heathland on the tithe award map of 1842 (NRO DN/TA 186) and, in some cases, on the Ordnance Survey second edition 6-inch map of 1906. Grigor described how the woods to the west of the river featured 'numbers of fine old veteran pollards standing singly, with their huge branches feathering to the ground, with fern, thorns, and brushwood, luxuriantly growing in the intervening spaces' (Grigor 1841, 232). Further concentrations of oak pollards are found on the Letheringsett estate, some 1.5km to the east, again buried within an eighteenth-century plantation ('Pereer's Hills'); and 1km to the south, within an eighteenth-century plantation called Sand Hill, on the parish boundary between Saxlingham and Thornage, where there are around fifty oak pollards. In general, the trees are not large – almost all are less than 5m in circumference,

FIGURE 49. The Bayfield Oak, one of several ancient pollards that now grow within eighteenth- and nineteenth-century woodland on the Bayfield estate, but which originated as a tract of heathland wood-pasture.

and a number less than 3m – although here, as at the other places just noted, the poor, acidic and well-drained soils and the densely spaced character of the planting may well have militated against normal rates of growth. It is noticeable that the majority of the pollards are found on steeply sloping ground, perhaps suggesting that they survived better here because the intensity of grazing was less and sporadic cultivation of the heaths impossible.

Another important concentration of ancient pollards exists on the Greensand ridge to the north of King's Lynn in west Norfolk. Buried within the eighteenth- and nineteenth-century woodland around Ken Hill House near Snettisham, surrounding an area of surviving heath, are a number of oaks with girths of between *c.*4.5m and 5.5m. The area is shown as 'Caen Wood' on Faden's map of 1797, but appears to have been heathland in the early seventeenth century and by the time of the parliamentary enclosure of the parish in 1766 had partly been enclosed as a rabbit warren (NRO Le Strange OB2; NRO BO1). Further examples can be found, perhaps less surprisingly, on the islands of former heathland scattered across the clay plateau in the centre and south of the county. Substantial groups of ancient pollards, again all oaks, thus exist within Thursford Wood, a nature reserve in north Norfolk managed by the Norfolk Wildlife Trust which partly overlies sands and gravels: and in Middle Heath Plantation, some 2km to the north-east, which lies entirely on sandy and gravelly soils and which, before enclosure in the early nineteenth century, formed the northern section of Stock Heath, an extensive tract of common land shared by a number of parishes in the area. These trees were also noted by Grigor, who described how 'some hundreds of acres are here thickly strewed with thorns and holly of most magnificent growth, and ever and anon an old and gnarled oak contrasts and enlivens the scenery' (Grigor 1841, 233). At Middle Heath Plantation, once again, few of the trees are large: all but two of the thirty recorded here have girths of less than 5m. Large trees are better represented at Thursford, with one particularly massive example of 7.7m, but even here around three-quarters of the thirty recorded specimens have girths of less than 5m (Figure 50).

A number of relict wood-pasture sites are also known from Breckland. Broom Covert in Quidenham is a small L-shaped plantation presumably planted in the eighteenth century and already shown on Faden's county map of 1797. Within it, growing in two main groups, are thirty pollarded oaks with girths of between 2.5m and 5.0m (average 3.9m). The wood contains earthworks of boundary banks and a hollow-way, possibly suggesting that this was an area of private rather than common wood-pasture. Once again the small size of the trees is noteworthy. It is likely that their growth has been impaired by competition with the dense vegetation around and above them (many, indeed, are now dead or dying), as well as by the poor quality of the soils. Yet, even allowing for these things, it is difficult to believe that the smallest examples – with girths of 3m

FIGURE 50. These oak pollards, hidden away within Middle Heath Plantation in Thursford, grew on the northern portion of Stock Heath before its enclosure in the early nineteenth century.

or less – can have been very old when engulfed by the plantation, perhaps in the mid-eighteenth century. Here, as in the other cases where fragments of wood-pasture are preserved within later estate woodland, it is possible that the primary use of the latter as game cover (indicated in this particular case by the name 'Covert') may have encouraged the retention of existing trees because of the cover they afforded for game birds. Where plantations were established on former heaths primarily with profit in mind, existing pollards would perhaps have been removed more systematically. Small groups of pollards surviving within Grenadier Plantation in Garboldisham and in the woods forming the western edge of Merton Park may represent the remains of wood-pastures rather than former hedgerow trees, but more striking is the concentration of sixteen old pollards in the woods and scrubland around Stow Bedon Mere, an area which was still open heathland when Faden's county map was surveyed in the 1790s. Two-thirds have girths of less than 5m, but the others include one magnificent 'champion' tree with a circumference of around 8m. It is noteworthy that all these Breckland sites are found towards the margins of the region, on acid soils of the Newport 4 or Ollerton Associations – soils formed in gravels as well as sands, and in the case of the latter affected to some extent by a seasonally high water table. These are very similar to the soils on which the heathland wood-pastures occur in the north and centre of the county. The centre of Breckland, in contrast, is characterised by sandier soils overlying chalk at varying depths. Such soils were more attractive to early agriculturalists and here, as in the conventional story, trees may well have disappeared wholesale at a very early date. Large tracts of central Breckland were also given over to rabbit farms in the Middle Ages, further militating against the survival of trees; while the more calcareous ground was often cultivated at intervals (Bailey 1989). Not all medieval heathland in the county may thus have been the same. Where acid soils included a significant gravel component and/or had a high water table wood-pastures were perhaps more likely to survive than where they were drier and formed entirely in sand over chalk (Barnes *et al.* 2007).

The Felbrigg beeches

The pollarded beeches found on the acid soils within and around Felbrigg Great Wood in north Norfolk have been discussed on a number of occasions, and it has been suggested that they are among the most northerly indigenous examples of this species to be found in England (Figure 51) (Rackham 1986a, 141; 1976, 27). The soils in the area would normally have carried heathland, and a sketch map of 1777 describes the area occupied by one of the main concentrations of pollards at Felbrigg specifically as 'heath' (NRO C/Sce 2 Road Order Box no. 21). Eighty-one of the beeches recorded in and around Felbrigg have girths of between 4.0m and 4.9m, not large by the standards of this species and – to judge from dated examples – not necessarily any earlier than the early nineteenth century (above, pp. 52–4). A further fifty specimens have girths of between

FIGURE 51. The origins of the pollarded beeches at Felbrigg remain uncertain. They may be relics of the natural vegetation, but might have been planted on the Felbrigg estate by the Windham family in the late seventeenth century.

5.0m and 5.9m, but again – on analogy with other dated examples – there is little reason to believe that these are any older than the middle decades of the eighteenth century. Eleven are more substantial, with girths of 6m or more, but only three, with girths of 6.9m, 7.4m and 8.9m, are larger than an example of 6.8m in the avenue at Houghton known to have been planted in the 1730s. It is thus by no means impossible that none of the Felbrigg trees was planted before the later seventeenth century.

Heathland wood-pastures certainly existed at Felbrigg in the Middle Ages. In the 1490s William Hamund was granted the right to take 'all the underwood and lop all the trees that grow on the land of the said cottage and upon the separate common (*communam separalem*) opposite' (NRO WKC2/115). But it is possible that the Felbrigg beeches are a relatively recent addition to the landscape. William Windham of Felbrigg Hall began a sustained campaign of tree-planting in and around the park in *c.*1676, and it could be argued that, rather than being a managed relic of the native vegetation, the beeches represent another part of his estate planting, using stock brought in from elsewhere in England. It is noteworthy that James Grigor, although he provided a detailed description of the trees at Felbrigg, failed to comment on the extreme antiquity or size of any beech trees here. If, as seems possible, none were much more than 175 years old at the time he was writing, this is perhaps unsurprising. It is noteworthy that few if any ancient beeches have been found beyond the bounds of the Felbrigg estate, although the area of acid soils with which they are associated is rather more extensive. Nevertheless, the antiquity of the Felbrigg beeches, and the status of beech more generally in north Norfolk, remains unresolved. In particular, there are apparent references to beech standards in Edgefield Great Wood, some 10km to the south-west of Felbrigg, in the 1670s (Rackham 1986b, 326); beech pollards are mentioned in a survey of Blickling made in 1756 (NRO MC3/252), while beech pollen was recorded, albeit at low levels, in late prehistoric pollen cores taken from Diss Mere in south Norfolk. The question is evidently a complex one, which may be answered by further research.

The end of wood-pastures

Many heaths and other commons in Norfolk thus appear to have maintained a significant covering of trees well into the post-medieval period and are shown as wooded or partly wooded on a number of seventeenth-century maps; but it is likely that this woodland became increasingly open in the course of the seventeenth and eighteenth centuries. Faden's county map of 1797 clearly omitted some areas of wooded commons – as at the northern end of Stock Heath, already discussed. But it is surely significant that the majority are shown completely without trees, and that many of these were unquestionably wooded in the previous century, such as the significantly named Hook Wood Common in Morley, still shown as largely covered in trees on Thomas Waterman's map of 1629 (NRO PD3/111); or the commons at Gressenhall, likewise mapped by Waterman in the 1620s (Figure 52).

In most cases, any trees that survived on common land until the end of the eighteenth century were completely destroyed when the commons themselves were enclosed and divided by parliamentary acts. Commons like Fritton were rare: most Norfolk examples disappeared during the high price years of the Napoleonic Wars, when enclosure and reclamation reached a frenetic peak. Statements of Claim made in association with particular enclosures often make references to the trees growing on common land. That for the mid-Norfolk parish of Shipdham, for example, records that many of the claimants had the rights to a 'planting' on the common (on which Faden's map of 1797 shows a large number of trees growing). John Platfoot held three communable messuages and, along with rights to common pasture, to cut flags and furze for fuel and to excavate clay for repairs, he also claimed 'the planting of trees standing upon the said common pasture, opposite and adjoining the said premises respectively' (NRO BR 90/14/2). The Earl of Leicester, Lord of the Manor of Shipdham, claimed in respect of five messuages for:

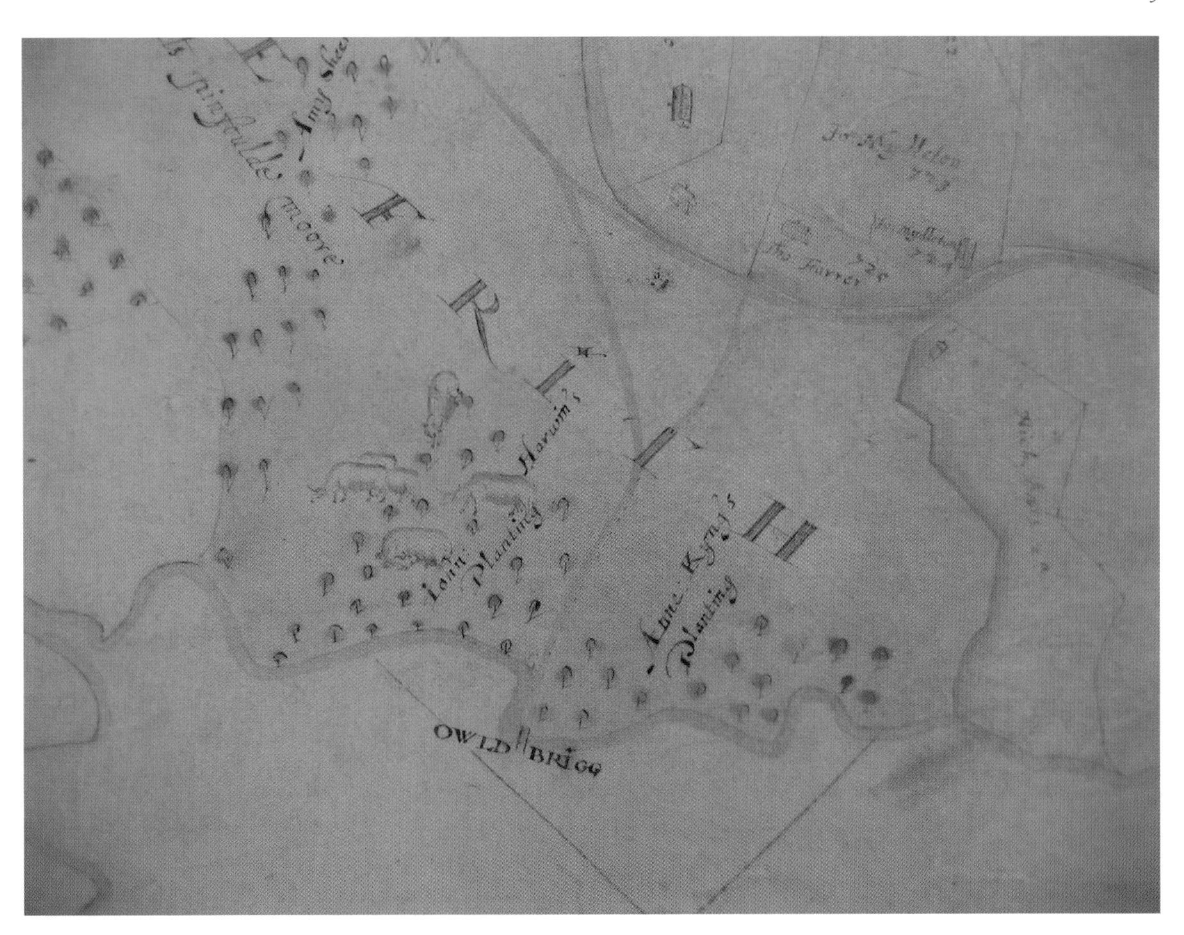

FIGURE 52. A wood-pasture common shown on Thomas Waterman's 1624 map of Gressenhall. Note how the areas of trees are described as 'plantings' and associated by name with particular tenants of the manor. (NRO Hayes and Storr 72)

> All trees, and all bushes and thorns planted or set by him or his predecessors, or his or their tenants, upon the said commons and waste grounds, contiguous or near to any of his said messuages or farms, which have been usually lopped, topped, pruned, or cut by him or his predecessors, or his or their tenants. (NRO BR 90/14/2)

Private wood-pastures, which were probably also a frequent feature of the landscape in the sixteenth and seventeenth centuries, also perished in the eighteenth and nineteenth. A number are shown on eighteenth-century maps, such as that surveyed by James Corbridge of lands in Guestwick in 1726 (NRO 723/3/11); and fragments survive in a number of places, as at Forncett (again buried within a later plantation) and Elsing, within the significantly named Old Pollard Grove (a few of the heathland wood-pastures already described may, as at Quidenham, have been on private land rather than commons). Most such areas had gone by the middle or later decades of the nineteenth century, victims, as Patsy Dallas has argued, of a fashion for 'improvement' which viewed such forms of land use as archaic and agriculturally inefficient (Dallas 2010a).

Conclusion

We have briefly outlined, over the previous pages, the history of traditional woodland management in Norfolk. In particular, we have described how the virgin 'wildwood' was extensively modified by several thousand years of human intervention before the eleventh and twelfth centuries; and how, under the impact of continued population growth, many of the remaining portions were then enclosed by manorial lords and managed more intensively as coppiced woodland or as deer parks or other forms of private wood-pasture. The rest continued as common land, but was further eroded by the expansion of cultivated ground and steadily became more open in character as the result of constant grazing pressure. Nevertheless, in a surprising number of places commons retained a significant number of trees well into the post-medieval period. This tells us something important about the nature of rural communities in the past, and especially about their relationship with the environment. Given the inherent instability of wood-pasture as a type of land use, the existence of so many wooded commons clearly indicates that communities had a greater ability to manage shared resources in a sustainable way than historians often assume.

Of particular note are the fragments of wood-pasture heaths which are now known from a number of places, perhaps most strikingly at Bayfield. It is noticeable that, in every case, these relic heathland wood-pastures survive within woods and plantations planted in the eighteenth or nineteenth century, a pattern that can be paralleled in a number of other parts of Britain, including Snowdonia (Edwards 1986) and the western Highlands of Scotland (Peterken 2008). Government agencies and a variety of conservation organisations are currently encouraging the restoration of heathland completely devoid of trees. The intention is to recreate uninterrupted swathes of heather, gorse and other classic heathland plants. But the discoveries outlined above suggest that heaths in Norfolk (and presumably elsewhere) often carried a significant covering of managed trees, interspersed with more open areas, and in some environmental contexts this, perhaps, is what we should be trying to recreate. Either way, it is important to emphasise that this recognition of the importance into modern times of wood-pasture as a form of land use and habitat rests not simply on the evidence of maps and documents, but also on that in the landscape itself, in the form, once again, of ancient, living trees.

CHAPTER FIVE

Meaning, Beauty and Commemoration

Introduction

So far we have been mainly concerned with the economic and agrarian role of trees in the past, with the ways in which they were managed as a source of wood, timber and fodder, and with how these demands, and the wider management of the farmed landscape, have affected the survival of old trees to the present. But in the past trees, and especially old trees, were also valued and often preserved for cultural reasons and were widely used as ornamental and commemorative features in parks, gardens and churchyards. Indeed, some social scientists have gone so far as to suggest that trees have 'agency' – that they have an active role in the structuring of social as well as environmental relationships (Jones and Cloke 2002): and while we would not go as far as this, it is clear that trees have always had an impact on human affairs going far beyond the simple economic.

Old trees were – like prehistoric barrows and other earthworks – ancient and immobile features of the landscape, and thus made excellent boundary markers. Not surprisingly, they often feature in descriptions of parish boundaries, the course of which were regularly walked and 'beaten' by local people not only in medieval times but in some cases into the nineteenth century (Winchester 1990). The precise line taken by such boundaries was a matter of keen interest for a range of practical reasons not only to the Church itself, which was keen to ascertain which parish particular pieces of land owed tithes to, but also to the wider community. From the sixteenth century the parish became the unit responsible for supporting the poor and indigent and for levying the resources needed for that support. 'Beating the bounds' was also an occasion for expressing communal solidarity against the claims of neighbouring parishes regarding such things as the rights to exploit particular areas of common land, as Nicola Whyte has demonstrated (Whyte 2008, 59–86).

A description of the perambulation of the bounds of Alburgh, written in 1794, describes in typical fashion how at one point the party went 'into the meadows up to the Great Oak'; while elsewhere they went 'left across the field to a Pollard Elm in the hedge' (NRO PD 196/82). Such trees were scored with marks, often in the form of a cross, to identify their status. A description of the parish boundary between Lamas and Little Hautbois in the east of the county mentions 'The Sallow Tree that is marked' and 'the Oak that is marked against the Wroxham lane' (NRO BR90/22/16). It is likely that a role as a boundary marker would preserve trees in situations in which they might otherwise have been felled. A map of Diss, surveyed in 1727, thus shows a single tree standing in the middle of a field, apparently left when two adjacent closes were thrown together and the intervening hedge removed; the 'perambulation way' between Diss and Roydon is shown running up to it and then changing direction (NRO MS 4525).

Even when such prominent marker trees were removed, and perhaps replaced by some other feature (a stone or a post), a memory of their presence might be retained for a while. A late-sixteenth-century map of Cawston thus marks, without explanation, 'the place where the elme stood'. A few years earlier a witness in a court case described how the people of the parish had 'gone ther pambulacon unto the said elme and so back againe through the said lane'. However, he also described how the 'ould decaied elme wch elme was the bound betweene Heverland [Haveringland] and Cawston' was a serious obstacle in the lane, and had therefore been cut down and replaced with a 'dole stone sett and placed in the stead thereof' (Whyte 2008, 72–3).

It is sometimes suggested that the importance of old trees, especially pollards, as boundary markers encouraged their preservation, and that a high proportion of veteran trees are thus to be found on parish boundaries. It is true that a significant number of veteran trees recorded in the survey – around 5 per cent – are so located, but this is only slightly more than might be expected by chance, and the difference can probably be explained by the fact that hedges, and hedgerow trees, on parish boundaries were more likely to survive the large-scale process of field amalgamation in the later twentieth century than were ordinary field boundaries, not least because parish bounds frequently coincided with the boundaries between farms or estates. Certainly, there is no evidence that veteran trees found in such locations are particularly old or large. Indeed, the range of recorded girths of such trees appears no different to that of the population of old and traditionally managed trees in Norfolk as a whole.

The character of parkland planting

In Norfolk, as elsewhere in England, many of our oldest trees are to be found in the parks laid out around great mansions and manor houses. It is often stated, or assumed, that such trees were always *planted* as parkland timber in the remote past. At several Norfolk country houses, such as Kimberley, ancient oaks are said to have been planted by historical figures such as Queen Elizabeth. In reality, veteran trees in parks have rather varied origins, and to understand their true character we must first say something about the more general history of the county's parklands.

Most Norfolk parks do not date back to the Middle Ages, or even to the period before 1700. 'Park' itself is a complex term which has changed its meaning over the centuries. The earliest parks were, as we have already explained, enclosures made from the dwindling 'wastes' in the early Middle

Ages in order to provide manorial lords with a source of wood and timber and a place to keep deer. They were private wood-pastures which functioned as venison farms and hunting grounds. They were a common feature of the medieval landscape – more than ninety examples are known, or suspected, at various places in the county, although many of these were small or short-lived (Yaxley 2005). Some, such as that at New Buckenham, seem to have formed part of the 'landscapes of lordship' laid out around the greatest residences, castles and palaces – elaborate displays of status which complemented the appearance of a noble residence (Liddiard 2000). But the majority lay in relatively remote places, some way from the homes of their owners, largely because they had been enclosed from the residual areas of woodland surviving towards the edges of the cleared and cultivated land. Many contained a 'lodge', often moated, which served as permanent accommodation for the keeper charged with maintaining the park and its deer and as temporary accommodation for hunting parties (Liddiard 2007). Surrounded, like coppiced woods, by a substantial earthwork bank and fence, parks were powerful symbols of lordly appropriation of the remaining areas of wild land. They were a particular feature of the more marginal soils where significant tracts of woodland had survived into the Middle Ages. In Norfolk they were thus mainly found on the heavier clays in the centre and south of the county, although they were also a feature of some areas of acid sands, especially on the outwash gravels to the north of Norwich and the Greensand ridge running north from King's Lynn (Yaxley 2005).

Few of the deer parks created in Norfolk in the period before *c.*1450 survived into the eighteenth century. Even those which flourished in the sixteenth and seventeenth centuries often failed to survive beyond 1700. There are a number of reasons for this general pattern of discontinuity. Firstly, from the fifteenth century deer parks began to decline in number, largely for economic reasons. Many were ploughed up or converted to open pastures or coppiced woodland by their owners. There were thus significantly fewer parks in early post-medieval Norfolk than there had been in the Middle Ages. Secondly, from the fifteenth century it became more normal for parks to be located in the immediate vicinity of great houses, rather than at a distance from them, and for them to be less densely wooded and more carefully designed in character, although they were still essentially wood-pastures, with a significant number of pollarded trees. Many old deer parks were thus gradually abandoned, and new ones created next to the owner's home, usually at the expense of farmed land. Lastly, over time the fortunes of particular families and particular estates waxed and waned. What were, in the fifteenth, sixteenth or seventeenth centuries, important residences, with deer parks attached, had often declined by the eighteenth century to minor manors or tenanted farms, and their parks destroyed accordingly (Dye 1990; Williamson 1998, 40–46). As a result of all these factors, very few of the oldest trees in Norfolk parks were planted there as parkland timber in medieval or early post-medieval times, simply because hardly any of these places had existed as parks then: most Norfolk parks were first created in the period after

1700. Even where there appears at first sight to be continuity – where a park is known to have existed in a certain parish both in the Middle Ages and in the eighteenth century, and where the same family has owned an estate throughout this time – this is often an illusion. A good example is Kimberley. The great medieval mansion of Wodehouse Towers stood on a moated site in Kimberley parish, some way to the west of the present mansion, which actually stands in Wymondham. The latter was erected in the early eighteenth century by Sir John Wodehouse, fourth Baronet. There was no continuity at all between the medieval park, which lay around the old moated site, and the early-eighteenth-century deer park created around Wodehouse's new mansion, which forms the core of the present, rather larger, landscape park (Taigel and Williamson 1991, 69–71).

In 1750 there were relatively few parks in Norfolk, perhaps no more than twenty-five. All of them contained deer, for a park was still, by definition, something you kept deer in. Around half had originated in the period before 1700, Blickling, Buckenham Tofts, Felbrigg, Melton Constable, Heydon, Raynham, Merton, Hedenham and Hunstanton among them. The others, including Langley, Earsham, Kimberley and Gunton, seem to have been created in the period between 1700 and 1750 (Williamson 1998). The overwhelming majority of surviving parks in the county were thus new creations of the later eighteenth or nineteenth centuries, laid out across farmland or over areas occupied by earlier formal gardens of varying size and complexity. Even where they had earlier origins they were significantly expanded in size during the eighteenth and nineteenth centuries, so that the original deer park forms only a small 'core' within the later park.

After 1750, and especially from *c.*1770, the number of parks in the county increased so rapidly that by the 1790s few mansions of any significance lacked one. Faden's map of Norfolk, published in 1797, depicts no less than 193 examples, although many were tiny affairs, covering less than ten hectares. These were 'landscape' parks – ornamental, naturalistic designs of open grass, scattered trees and clumps of woodland which were grazed mainly by sheep and cattle, rather than deer. They lacked the avenues and other kinds of geometric planting which had featured in many seventeenth-century deer parks, and were no longer usually managed systematically as wood-pastures, with large numbers of pollarded trees, for, as we have seen (above, p. 26), in the course of the eighteenth century landowners came to regard pollarding as an unsightly, backward, 'peasanty' practice, and one which should not be allowed to intrude into the landscapes of gentility. When parks of this new type were created, moreover, walled gardens and other enclosures around the mansion were removed, so that the park became its main setting. The new style is firmly, and to some extent rightly, associated in the popular mind with the name of Lancelot 'Capability' Brown. He was certainly the most successful of all the 'improvers', and was responsible for a number of important landscapes in the county, including Kimberley, Langley and Melton Constable. But there were other designers who ran national landscaping practices at this time: men like Nathaniel Richmond, who designed the parks at Beeston St Lawrence; or William Emes, who worked at Holkham in the 1780s (Brown 2001; Cowell 2006; Williamson 1998, 103, 126–7). Most parks, however, were the work of local surveyors or nurserymen (many of whom set up in business after having been employed by one of the more famous 'improvers'). New examples continued to appear well into the nineteenth century, especially in the vicinity of Norwich, while many existing parks continued to be embellished and expanded into the early years of the twentieth century.

Looked at in the long term, the history of parks in Norfolk (as in most other English counties) is thus characterised by a high degree of discontinuity, and for this reason it is hardly surprising that there are only a handful of places where trees planted in medieval or early post-medieval deer parks survive in the modern landscape. Some of the ancient oak pollards on the rising ground to the east of Bayfield Hall may once have stood within a deer park, to judge from the field names in the area, several of which feature the term 'Lowndes', perhaps from the medieval term 'laund', an open area within a deer park. The great Winfarthing Oak, already mentioned, seems originally to have grown within the medieval park at Gissing. Neither of these examples is now within a park, although the Bayfield examples lie on the fringes of one. Hunstanton Park in west Norfolk, which was created as a deer park in the fifteenth century (Oosteman 1994, 39; Dye 1990), does contain some ancient pollards of sixteenth- and seventeenth-century date, and some possible medieval examples, while the massive oaks at Merton and some of those at Heydon almost certainly first grew in the diminutive sixteenth- or seventeenth-century parks at those places (Williamson 1998, 241, 264). Of particular note are the oaks growing within Hedenham Park in the south-east of the county, a park which originated as a small deer park probably in the late seventeenth century. Here, many of the trees have been pollarded at a considerable height, in some cases as much as 4m above the ground, and it is possible that this was to keep the regrowth out of reach of deer. But these places are exceptions. Most ancient trees in Norfolk parks, of which there are many, have other origins. They are either survivals from the agricultural landscape which the park replaced or they are the remains of formal planting schemes – avenues and the like – laid out around mansions in the seventeenth or early eighteenth century.

Early formal planting

In Norfolk, as elsewhere in England, elite gardens in the period before the early eighteenth century were generally of a formal, structured character and were usually enclosed in whole or part by walls or fences. They featured knots or parterres (geometric patterns of box, grass and gravel), topiary and gravel walks, as well as terraces and elaborate garden buildings (Strong 1979; Laird 1992; Taigel and Williamson 1991). Several such gardens appear on early maps and illustrations, but little remains on the ground today, even of their 'hard' landscaping. This is because these kinds of garden were generally swept away from fashionable facades in the eighteenth century as the taste for simpler, 'naturalistic' styles of landscaping took hold, and the remains of only a few survive in Norfolk today, at places like Besthorpe or Intwood, generally where estates were downwardly mobile and absorbed into the property of some wealthier neighbour.

By the later seventeenth century it was common for geometric arrangements of trees to be extended out for some distance from the immediate vicinity of the house in the form of avenues framing distant views and vistas. These sometimes ran across adjacent areas of parkland, but more usually through the surrounding fields. Formal planting, like walled gardens, became unfashionable in the course of the eighteenth century, but more gradually. Avenues were still being established, even at major residences like Houghton or Holkham, into the 1730s and 40s, and simplified forms of geometric planting, involving rigid vistas framed by rectangular blocks of trees, might be created even later than this, long after walled enclosures had fallen from favour among the fashionable (Williamson 1998, 46–90). Although in most cases such arrangements were removed as the rage for 'natural' landscapes reached its peak in the later eighteenth century, this was by no means always the case. Some landowners retained elements of the old style of planting, perhaps because of the associations of dynastic longevity it proclaimed, or simply because they liked them. And once again, in some cases fragments survived because the associated mansion ceased to be the residence of an individual of wealth and taste. Whatever the explanation in particular cases, some of the most impressive veteran trees in Norfolk are survivals from the great geometric planting schemes of the period before *c.*1740.

Most of those planting schemes took the form of avenues. It is sometimes suggested that these were a French fashion, part of a more general influx of French styles into England in the late seventeenth century, and that few examples existed in the country before the English Civil War. But avenues were an obvious method of enhancing the views towards and away from a mansion, and of expressing power and ownership, for only absolute control over the surrounding land allowed an owner to plant across it. While it is true that avenues are rarely depicted on the small number of maps and illustrations which we have dating to the sixteenth or early seventeenth century, there is little doubt that some examples were established in this period. Local tradition holds that the sweet chestnut avenue immediately to the north of Heydon Hall was planted before 1600, perhaps when the hall itself was rebuilt in the 1580s, and the size and character of its few remaining trees certainly makes this possible (Figure 53). It must be emphasised, however, that most surviving avenues in the county are not very old at all, for this form of planting returned to fashion once again in the course of the nineteenth century. And even where avenues did originate in the seventeenth or early eighteenth century many – often most – of their original trees have been replaced, in part because avenues are particularly susceptible to gale damage.

FIGURE 53. The remains of the sweet chestnut avenue, possibly planted before 1600, to the north of Heydon Hall in north Norfolk.

Lime was particularly favoured as an avenue tree, probably because of its relatively rapid growth combined with its graceful form when young. The great avenue at Rougham, planted soon after 1693 by Roger North, still runs for some 750m to the south of the site of the hall (which was itself demolished in the eighteenth century, although the surrounding parkland survives) (Williamson 1996a). It was originally more than twice this length. Although most of the surviving trees are replacements, some of the originals, with girths of between 5.4m and 5.8m, are still thriving. At Merton, three huge lime trees – with girths ranging from 8.3m to 8.8m – in the area to the north of the hall are the remains of the avenue shown here on an estate map of 1733 (NRO NNRO 90/2 microfilm), and which was probably of late-seventeenth-century date (Figure 54). An equally popular choice as an avenue tree was sweet chestnut. The four massive survivors of the possibly sixteenth-century avenue at Heydon Hall, a feature already shown as in decline on an estate map of 1776 (NRO 334/1,3 microfilm), have girths of 5.1m, 7.1m, 7.6m and 8.5m. Equally striking is the great avenue running north-east from Houghton Hall, the only survivor from a more extensive and complex mesh of geometric planting which was largely removed in the 1730s when the landscape of the park was transformed by the great designer Charles Bridgeman (Williamson 1996b, 44–6). Here, the largest of the original trees (remaining within much later replanting) have girths of over 7m (Figure 55). Some of the trees which made up the other avenues at Houghton were also allowed to remain, now as scattered free-standing specimens within parkland: and these survivors suggest that most if not all of the avenues swept away by Bridgeman were likewise planted with this species.

Bridgeman's new design for the park at Houghton, typically for the period, was still essentially formal and geometric, but simpler and less cluttered in

FIGURE 54. Huge lime trees, the remains of a seventeenth-century avenue, in the park at Merton on the edge of Breckland.

FIGURE 55. A sweet chestnut avenue at Houghton, the only survivor of a dense mesh planted in the park here in the seventeenth century. The others were felled when the landscape was redesigned by Charles Bridgeman in *c.*1730.

character, with four main avenues and vistas focused on the hall. The planting was also different. The avenue running south from the hall, the 'South View', was composed of double-planted oaks: some of the original trees survive, with girths of 3.2m to 5.0m, among much later replanting. That to the north – the 'North View' – was composed of beech, and again a small number of original trees remain, with girths from 4.8m to as much as 6.8m. In general, oak, beech and elm seem to have been more widely planted in geometric schemes in the early eighteenth century, although lime and sweet chestnut continued to be used extensively. At Wolterton, for example, lines of sweet chestnut were planted in the 1720s to frame the rectangular lawn lying to the north of the hall. Some of these still remain, massive specimens with girths of between 6m and 8m (Williamson 1998, 286–8).

As well as avenues running across parks and estate land, the formal designs of the seventeenth and early eighteenth centuries also featured extensive areas of ornamental woodland and shrubbery, called 'wildernesses' and 'groves'. These were dissected by paths, usually hedged, which were initially straight (and often laid out in the form of a St Andrews Cross) but which, in the 1720s and 30s, tended to become more serpentine and curvilinear in character. Wildernesses are often referred to in the documentary record, but only fragments usually survive in the landscape today. Their planting was often complex. That at Raynham, for example, described in an undated contract of *c.*1700, had paths lined with hedges of hornbeam running through woodland planted with spruce, 'silver fir', lime, horse chestnut, wild service, beech, sycamore and birch, underplanted with laburnum, guelder rose, lilac and other flowering shrubs (Raynham Hall archives: Williamson 1998, 269). No trace of this feature survives today, and elsewhere only the longest-lived trees now remain to mark where these complex areas of planting once existed. At Rougham what is presumably the remains of a wilderness of some kind, ranged either side of the great lime avenue immediately to the south of the site of the hall, contains more than thirty massive, gnarled sweet chestnuts, with girths ranging from 5m to 7m, arranged in a complex fan-like pattern. Large old sweet chestnuts also survive among much later replanting immediately to the west of Wolterton, within the area shown as a wilderness on a map of 1732 (and perhaps planted when this area was reorganised in the mid-1730s) (Wolterton Hall archives, Wolt 10/120; Wolt 8/12). Large trees of the same species (with girths of 6.1m, 7.0m and 7.2m) in the park at Great Melton, to the east of the site of the hall, may likewise represent the remains of a wilderness or similar feature, left to grow on as free-standing parkland specimens when the feature itself was removed.

One of the best surviving wildernesses in the county is also one of the latest – that immediately to the north of Gunton Hall, probably designed once again by Charles Bridgeman shortly before his death in 1739, and featuring a huge circular viewing mount (Willis 1977, 85, 132). It is difficult to distinguish original trees from later replanting, but the former include some massive examples of oak and sweet chestnut, together with a few of the beech trees which originally lined the complex pattern of paths. Later wildernesses were occasionally placed at a greater distance from the house, as at Holkham, where Obelisk Wood, placed on the hill over 500m to the south of the house, was laid out in the 1720s with a mesh of straight paths, some focused on particular features in the surrounding landscape, such as church towers (Holkham Hall archives a/32; NRO Ms 21127a, 179X4)). Beech was one of the main trees planted but none have survived in this dry and exposed situation. However, some of the holm oaks still growing here may be original, although most – and the many examples more widely scattered through the park – are not (the tree, favoured for its Italianate associations, does well by the coast, thrives in the park, and was continually planted here right through the nineteenth century).

There are other relics of early formal planting surviving in the county, including the great limes at Kimberley, with girths of around 7m, which originally formed part of rectangular blocks of planting framing the vista to the south-west of the hall; fragments of the avenue at Docking, planted in the early eighteenth century, planted with lime and beech (the survivors now with girths of around 6m); and in the great avenue ('The Chase') running north from Stow Bardolph Hall, where there are oaks with girths of around 5m (in the northern section) and sweet chestnuts with girths of just over 5m (in the southern). Several large individual trees in country house gardens and pleasure grounds almost certainly once formed part of avenues, wildernesses or other geometric planting, most notably the massive sweet chestnut (with a girth of 8.3m) in the grounds of Hanworth Hall.

Parks and pleasure grounds

As we have already emphasised on a number of occasions, the oldest trees found in Norfolk's parks are generally survivors from the old, pre-park landscape. It made sense to keep a proportion of hedgerow trees when the hedges themselves were removed to create the open, uninterrupted vistas expected in parkland. Most were pollards but, once incorporated into landscapes of display and leisure, they ceased to be cropped, and were allowed to grow on as naturally as they could, although retaining their short trunk or bolling. Wholesale retention of older trees in this way made a new park look like a long-established one: but not all contemporaries were fooled. William Gilpin perceptively noted in the 1760s how, around Robert Walpole's Houghton Hall in Norfolk, 'it is easy to trace, from the growth of the woods, and the vestiges of hedge-rows, where the ambition of the minister made his ornamental inroads into the acres of his inheritance' (Gilpin 1809, 41). Almost all such trees are oaks: examples of particular note include Kett's Oak in Ryston Park, a huge specimen with a girth of nearly 9m; the great oak, with a circumference of slightly under 9m, associated with the site of a deserted medieval settlement in the north of Houghton Park (Figure 56); and the remarkable collection of old pollards at Kimberley (Figure 57). Other notable concentrations occur in the parks

FIGURE 56. A massive oak tree, possibly a former shred, in Houghton Park. The tree grows on an earthwork which marks a former boundary within a part of Houghton village, which probably disappeared in the fifteenth century.

FIGURE 57. Ancient oak pollards, some of medieval date, flank a 'hollow-way' or former road within the park at Kimberley. The park contains one of the most important collections of veteran oak trees in Norfolk.

at Heydon, Blickling, Weston Longville and Hedenham. Such ancient trees are often associated with the earthworks of old field boundaries or the linear depressions called 'hollow-ways' marking the line of former roads usually closed to the public when the park was created. Trees and earthworks form, as it were, parts of the same archaeological landscape. Particularly striking in this respect is the fine park at Raveningham, laid out in the 1780s, where pre-park pollards with girths ranging from 3.6m to as much as 7.2m are associated with old earthwork banks which can be correlated with fields and roads shown on an early-seventeenth-century estate map (Raveningham Hall archives).

It is noteworthy that the most ancient trees were not normally retained close to the mansion; nor did they generally survive where carefully composed views and vistas were created. Most are to be found in the more distant recesses of the park, where they seem to have served as interesting destinations for a walk or ride. At Houghton, for example, the oldest oaks are to be found some way from the hall, in the north park, well away from the more overtly 'designed' parts of the grounds. Kett's Oak in Ryston Park is, similarly, located over 600m to the south of the hall. Some examples, however, are found close to main entrance drives, perhaps providing visitors with an initial and abiding impression of the longevity of the landscape, and thus of the family that owned it. At Kimberley, for example, most of the ancient oaks are to be found to either side of the main approach; none have been allowed to remain in the principal view, designed by 'Capability' Brown himself, extending from the house westwards to the great lake.

In some landscape parks pre-park timber, usually oak pollards, is particularly prominent, occasionally forming the majority of the trees that are present. This is often where the park in question has had a chequered history in the twentieth century, passing though a number of hands. In such circumstances the more valuable timber – the standard trees – has often been felled and sold, leaving the old pollards, which were of little interest to timber merchants. Most of the trees in Catton Park, for example, are former hedgerow pollards, but a century ago they would unquestionably have been outnumbered by other trees planted when the park was first created or added at some subsequent date.

Eighteenth- and early-nineteenth-century parks in East Anglia thus contain three broad categories of tree: timber retained from earlier geometric planting, usually suitably thinned; trees retained from the agricultural landscape which preceded the park, mainly hedgerow timber; and new trees, planted as part of the initial design or at some subsequent date, either as free-standing specimens or in clumps or larger blocks of woodland. Only a small proportion of the trees in this last category are of sufficient size to have been considered 'veteran' by the various recorders involved in the surveys which form the basis for this book. Most free-standing oaks of later-eighteenth-century vintage, for example, have girths of less than 4m. But parkland limes planted in the period after 1760 can comfortably attain a girth of 5m, while some colossal specimens of beech were planted only in this period, including examples at Raveningham (5.0m and

5.2m), Docking (5.5m) and probably Bayfield (6.8m) (Figure 58). Cedars planted in parks and pleasure grounds in this period can likewise attain considerable girth, frequently in excess of 5m.

Late-eighteenth- and early-nineteenth-century mansions did not stand alone, 'solitary and unconnected', in open parkland. Although enclosed and geometric gardens were swept away by Capability Brown and his 'imitators', areas of more informal pleasure ground were created, usually to one side of the main facade, featuring winding paths, shrubs, flowers and specimen trees – a number of examples of which survive in Norfolk, some so large that they have effectively attained veteran status. Particularly striking are the London planes in the pleasure grounds at Ryston (with girths of 5.7m and 5.9m). At nearby Stow

FIGURE 58. In spite of its vast size, this magnificent beech beside Bayfield Hall was probably planted only in the middle decades of the eighteenth century.

Bardolph the grounds, laid out by Lewis Kennedy in 1812, still feature London planes with girths of more than 4.0m, beeches up to 5.0m, cedars of Lebanon with girths of 5.8m and a Deodar cedar of no less than 6.9m (Williamson 1998, 279–80).

To judge from surviving trees, and also to some extent from documentary evidence, the planting in mid- and later-eighteenth-century parks was dominated by oak, with lesser amounts of beech, sweet chestnut and lime. Some conifers (especially Scots pine) were also established, but only cedars seem to have survived to the present. From the 1790s, however, planting appears to have become more varied, with a higher proportion of lime and beech and increasing quantities of horse chestnut and London plane. The trend towards more varied planting intensified through the nineteenth century, elaborate formal gardens returned once more to fashion, and avenues appeared once again in parks. But very few trees planted around great mansions after *c.*1850 really qualify as 'veterans', although a number of horse chestnuts with girths of up to 5m may have been planted in the middle decades of the nineteenth century, and examples of cedar of Lebanon and Wellingtonia can appear ancient even if they are little more than a century and a half old. As formal geometric planting came back into favour particular species might be used in novel and striking ways (Elliott 1986). At Lynford in the Norfolk Breckland, for example, the great Victorian designer William Andrews Nesfield planted a magnificent avenue of Wellingtonias, now grown to vast size, only a few years after that species had been introduced from America (*Gardeners' Chronicle* 20 Sept 1884).

Churchyard trees

As well as finding trees aesthetically pleasing, people in the past – as today – have invested them with a range of symbolic meanings. They have used large and distinctive examples as memorials to important people or events, and have invested them with a spiritual significance. In pre-Christian times trees, together with other natural features, such as springs, seem to have had a real religious role as objects of veneration, as place-names such as Ashwellthorpe (originally Ashwell: 'the spring by the ash') may suggest. The remarkable site which has become known as 'Seahenge' shows that the veneration of trees extends back far into prehistory. This monument, dating from the Bronze Age and discovered on the foreshore at Thornham in 1998, comprised an inverted oak trunk, surrounded by a ring of split oak posts (Pryor 2002). In some other parts of England, principally the south and west of the country, a number of churchyards contain massive old yew trees which, it is claimed, are older than the church itself, and represent the pre-Christian focus for worship on the site (Bevan-Jones 2002; Howes 2009, 94–5). In Norfolk, and in eastern England more generally, there does not seem to be anything comparable to the great churchyard yews at Payhembury in Devon (with a girth of more than 15m) or Fortingall in Perthshire (nearly 18.5m in circumference) (Bevan-Jones 2002,

189, 191). The largest churchyard yews recorded in Norfolk do not exceed 5m in girth (although a few coppiced examples have stools as much as 6m in circumference). Most are normal English yews, but some are of the fastigiate Irish variety, introduced from Fermanagh in the 1780s but only widely planted from the early nineteenth century (Mitchell 1974, 52). Yews are notoriously difficult to age – they put on girth erratically – but it is perhaps noteworthy that several examples of Irish yew in Norfolk churchyards have girths in excess of 3m, perhaps suggesting that most if not all of the yews recorded in the county's churchyards are of eighteenth- or nineteenth-century date, a suggestion supported by the size of dated yews in other contexts. Several of those planted to screen the kitchen garden at Ditchingham, for example, unquestionably in the period after 1764, have girths well in excess of 4m.

As with other evergreens found in churchyards and cemeteries, the presence of yews may in part reflect the influence of John Claudius Loudon's important book of 1843, *On the Laying Out, Planting, and Managing of Cemeteries* (Loudon 1843), which encouraged a surge of planting in burial grounds of all kinds. Before this date, churchyards – in Norfolk at least – seem to have been only sparsely planted. James Grigor, writing in 1840, noted that 'In Norfolk the yew is very scarce, and in the churchyards almost unknown. We are miserably behind continental nations in the respect we pay to the decorating of the depositories of the dead. Some of our churchyards are absolute wastes – without a tree of any description' (Grigor 1841, 149). Similarly, Ladbrooke's engravings of Norfolk churches, published in 1843, generally show rather bare and open graveyards, devoid of trees and shrubs (Ladbrooke 1843). This said, there are stray documentary references to earlier planting in churchyards. The planting of oaks is recorded at Salhouse in 1746, for example, and that of walnuts at Briningham in 1750 (NRO PD 625/2; PD 646/1). Moreover, while most trees in churchyards appear to be of nineteenth- or twentieth-century date, there are a number of exceptions. These include a handful of large beech trees, including a specimen with a girth of 5.25m at North Tuddenham and one reaching 6.0m at Walsoken; a number of sycamores with girths greater than 5m (at Stow Bardolph, Taverham, Kirstead); limes with girths of more than 5m (Marsham, South Raynham, Kelling) and holm oaks of more than 6m (at Ryston and Anmer). The massive sweet chestnut at Hevingham, reputedly planted around 1647 and with a girth of around 7.5m, has already been discussed: examples of 6.1m at North Burlingham and of 6.6m at Barton Turf are also likely to have been planted in the seventeenth or early eighteenth century.

Most trees in Norfolk churchyards are thus of nineteenth or twentieth-century date, but a small minority are older. These, however, appear to be decorative or even practical rather than symbolic or commemorative in character, little different from the kinds of trees that might be found in the grounds of rectories and vicarages, and were planted by incumbents who were, in many cases, the younger brothers of the men who supervised the more extensive plantings of country house parks and pleasure grounds. There are certainly no examples in

FIGURE 59. The Hethel Thorn, now protected as a diminutive nature reserve managed by the Norfolk Wildlife Trust, grows in a field to the north of Hethel church in south Norfolk.

the county of the kind of ancient, arguably pre-Christian, trees found elsewhere in England, although in this context mention should perhaps be made of the 'Hethel Thorn', which stands in a field just to the north of the parish church of Hethel, rather than in the churchyard itself (Figure 59). It is a large hawthorn, the main trunk of which is split into two sections with circumferences of 1.1m and 1.5m respectively. It stands on a low earthwork of uncertain age or significance. When Grigor described the tree it was much larger, with a trunk over twelve feet (*c.*3.7m) in circumference: this later disintegrated, leaving the less impressive specimen we see here today. Although there is no way of checking the various claims which have been made for the tree's extreme antiquity (anything up to a thousand years) it was already being noted as a massive and ancient specimen in the 1750s, and according to local legend had grown from the staff of Joseph of Arimethea. The thorn is one of the oldest in England, and forms one of the country's smallest nature reserves. It seems an obvious candidate for a tree with a pre-Christian religious significance, but, although unquestionably very old, it seems unlikely that it dates back to the pre-Christian era (well over a thousand years ago) and, as noted, it does not actually grown within the churchyard itself. Other stories were told about it in the past. In the middle years of the nineteenth century it was also known as 'the Witch of Hethel', while Grigor reported claims that 'in one of the chronicles, the thorn was mentioned as the mark for meeting in an insurrection of the peasants in the reign of King John'. He added that he had 'never been able to get a reference to what chronicle' (Grigor 1841, 283).

Celebration and preservation

The *Gentleman's Magazine* for 1856 described how the Hethel Thorn was being 'carefully protected from injury by a fence maintained by direction of Mr Hudson Gurney' (*Gentleman's Magazine* 1856, 763). It was clearly one of several ancient trees in the county which were widely visited and written about in the eighteenth and nineteenth centuries, part of a more general interest in arboriculture among members of the landed gentry, many of whom were involved in sustained campaigns of tree-planting on their estates. Robert Marsham, the great Norfolk naturalist and father of the science of phenology, had an obsessive interest in trees and undertook a systematic programme of afforestation on his properties at Stratton Strawless, to the north of Norwich, in the middle and later decades of the eighteenth century. From 1734 he carefully recorded the increases in the girths of the trees he had planted, communicating this information at regular intervals to the Royal Society and publishing in its journal an account of the various experiments he undertook into stimulating the growth of trees, which included (famously) the practice of washing and scrubbing their trunks (Ketton-Cremer 1957, 152–7). Such enthusiasms shaded easily into an interest in old large trees, although such an interest was also, in the later eighteenth century, stimulated by a growing interest in the picturesque. Perhaps the most striking manifestation of this is the remarkable book by James Grigor, published in 1841, to which we have already had cause to refer on numerous occasions: *The Eastern Arboretum, or Register of Remarkable Trees … in the County of Norfolk*. Grigor, a nurseryman by profession, was interested in all fine trees in Norfolk, especially rare and ornamental examples in the grounds of the county's country houses. But he included, as we have described, many references to particularly old or massive specimens, both in parks and gardens and in the wider countryside.

It is noteworthy that many of these large old trees had specific names, such as the 'King of Thorpe', with a girth of 6.6m; the 'King of Sprowston', with a circumference of 4.6m; and the 'Kempston Oak', which Grigor believed to be around 400 years old, 'whose remarkable appearance frequently induces the passing traveller to step out of his way to gaze with mingled feelings of wonder and admiration on its venerable form' (Grigor 1841, 338). He also described the 'Seven Brothers', a massive lime tree in Sprowston with seven prominent limbs (Grigor 1841, 201); the elm called the 'Deopham High Tree' (Grigor 1841, 361); and the 'Old Bale Oak' which, according to the historian Blomefield, could hold a dozen people within its hollow trunk. Landmark trees like these were sometimes noted on maps. A map of 1756 marks 'The Great Oak' standing in Hillington Park in west Norfolk (a large old pollard, with a girth of 6.1m, stands on or close to this position today) (NRO NRS 21368). Some large trees had a national reputation, like the great lime tree at Deopham, a massive specimen, recorded by John Evelyn in his *Sylva* of 1664 (Evelyn 1664, 82), which fell in 1705.

Interest in such distinctive features of the landscape was not limited to an educated elite. Old and massive trees were a source of pride to local people. Grigor described how the 'King of Thorpe' had 'become to the villagers an object of veneration and awe: their children know it, and run eagerly and proudly to show it to us'. He was informed 'by those who live here, both with sorrow and joy, "that it has been three times condemned and as often reprieved"'. He recounted how the 'Marsham Oak', which stood beside the highway in Worstead, had been cut down some years before: 'Evil was the hour of its fall to the inhabitants of the surrounding neighbourhood, who beheld it with fond veneration' (Grigor 1841, 228). The felling of a great tree would be attended with the same level of enthusiasm as a public execution. When two trees at Broome, the Admiral and the Vice-admiral, were cut down 'Many thousands of the neighbourhood flocked to see their destined downfall.'

FIGURE 60. The Winfarthing Oak, the largest tree in nineteenth-century Norfolk, as depicted in James Grigor's *Eastern Arboretum* of 1841.

Local people went to some lengths both to protect old trees and to display them to visitors. Grigor reported how there was an elm tree at Acle 'which the inhabitants seem to delight in; it stands in the centre of the town, protected with wooden rails' (Grigor 1841, 363). The interior of the great Winfarthing Oak could be reached through a door (Figures 60 and 61), while a large elm (a former pollard) in the grounds of Salhouse Hall had a seat inside it. Local efforts to protect and display were not always held to be sufficient by educated visitors, however. Grigor commented, for example, on the Great Oak at Thorpe:

> Were this tree ours, we would immediately have a space of fifty yards cleared around it, and, after laying it down in turf, have it enclosed with iron rails or common palisading. We should, furthermore, have an elegant gravel walk leading to it from the village of Thorpe, in order that the public might visit it conveniently; for, at present, it is of most difficult access, being surrounded by woody undergrowths and rank herbage. (Grigor 1841, 134)

In a similar way, the great oak at Winfarthing could not be reached along a proper roadway, but only along a pathway 'intersected by a series of stiles': 'It is, besides, to be regretted, now that it cannot be seen amidst rough uncultured scenes, that no effort has been made to clear a sufficient space around, for the purpose of being laid down to turf and occupied with seats on which visitors might sit' (Grigor 1841, 355).

The Winfarthing Oak was perhaps the greatest and most famous of all named trees in nineteenth-century Norfolk, renowned well beyond the boundaries of the county. It is mentioned in both Loudon's *Arboretum Britannicum* of 1838 and Samuel Taylor's *Arboretum et Fruticetem Britannicum* of 1836 (Grigor 1841, 354; Amyot 1874), and is the only tree individually labelled on William Faden's county map of 1797. By the nineteenth century it was already completely hollow and propped up with assorted poles. A plaque proclaimed that, when measured in 1820, 'This oak in circumference at the extremities of the roots is 70 feet: in the middle, 40 feet.' Taylor described it as 'a mere shell, a mighty ruin, blasted to a snowy white, but it is magnificent in its decay'; Loudon believed that it was 1500 years old. In the nineteenth century it was also known as the Bible Oak, not because it stood on a perambulation way but rather because a collecting

FIGURE 61. The Winfarthing Oak and its companion tree, the 'Little Oak', as shown on the Tithe Award map of 1840. (NRO DN/TA 532)

OAK – WINFARTHING.

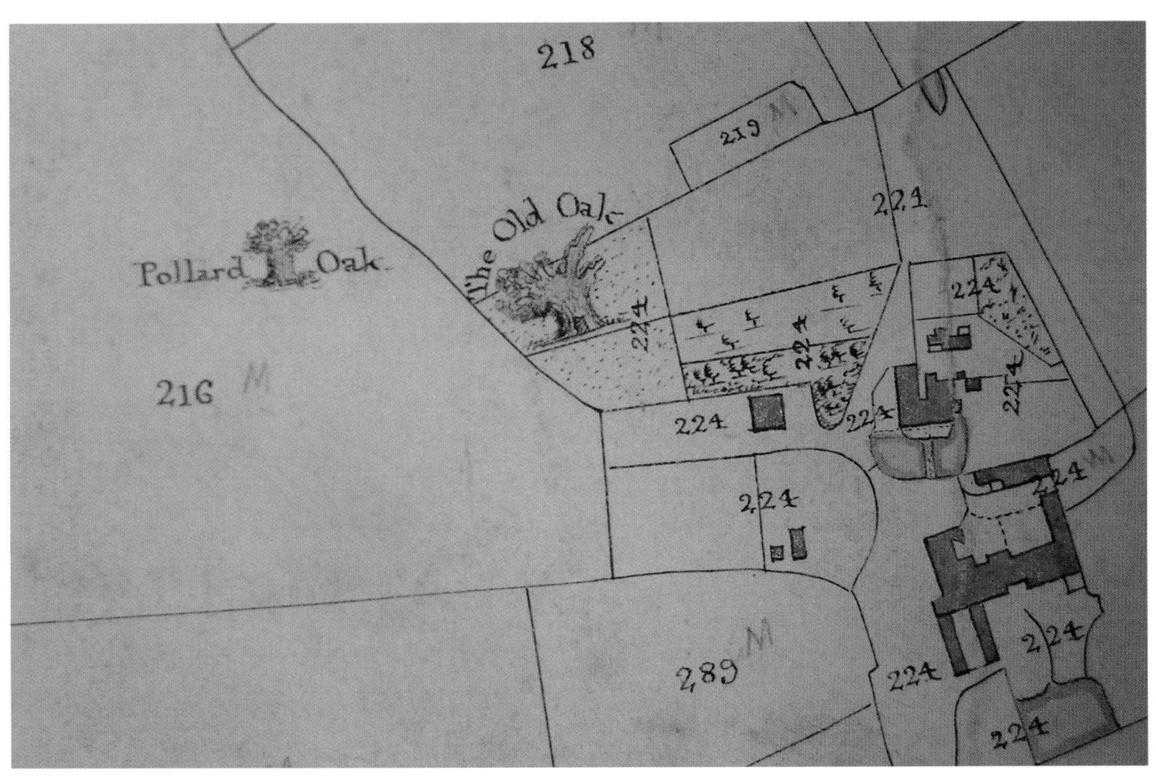
218
219
221
Pollard Oak
The Old Oak
224
216
289

box for the British and Foreign Bible Society had been attached to its side. The tree, together with its neighbour the 'smaller oak', is clearly shown on the Winfarthing tithe award map of 1838 (see Figure 61). Both had originally grown within the medieval deer park of Winfarthing (they stood next to 'Lodge Farm') and, as already noted, the larger tree at least was perhaps the last survivor of the great swine-woods described by Domesday Book. By the nineteenth century it was clearly in decline. Grigor described how the only life it exhibited was on the south side, where a narrow strip of bark set forth 'a few branches'. Nevertheless, when the tree was visited in 1894 by members of the Norfolk Naturalists' Trust it was still not quite dead (Amyot 1894): it finally died a few years later, although its shell remained recognisable into the 1950s.

Another well-known tree in nineteenth-century Norfolk was Kett's Oak in Hethersett, which still survives (Figure 62). This was traditionally said to have been the 'Oak of Reformation' at which the rebel Robert Kett preached to his supporters before they marched on Norwich in 1549. In 1829 it was described as 'an aged oak, whose trunk though wasted to a mere shell, still bears a crown of lively green' (Chambers 1829, 801). Grigor illustrated it (Figure 63), and measured its girth as seven feet seven inches (*c.*2.6m) at a height of five feet

FIGURE 62. Kett's Oak at Hethersett, reputedly the 'Oak of Reformation' where William Kett addressed his rebels in 1549.

FIGURE 63. Kett's Oak at Hethersett, as depicted in James Grigor's *Eastern Arboretum* of 1841. The small size of the tree then, as now, makes it hard to believe that it was already a mature specimen in the early sixteenth century.

above the ground (Grigor 1841, 326). This seems rather small for a tree which was by then supposedly over three centuries old, and it is possible that the association with Kett was made some time after the sixteenth century, or that the tree has taken over the associations of some older, long-lost neighbour – as happened in the case of the famous Boscobel Oak in Shropshire (Stamper 2002). Either way, the story was clearly well established by the eighteenth century, and by Grigor's time measures had already been made to preserve the tree by binding its splitting trunk with a strip of iron. It has since put on another full metre of girth, another clear indication that it is not of any great age. This is in spite of the fact that in 1933 it was subject to yet more drastic conservation measures, when rotten wood was scraped from its cavity, which was then painted with plastic bitumen and filled with concrete (Long 1933). Surprisingly, it continues to flourish. There is another Kett's Oak in the west of the county, in the grounds of Ryston Hall (Figure 64). 'It is highly grotesque in its outline, and we have seldom seen a tree with so much of the fearful in its character' (Grigor 1841, 348). Six of Kett's rebels were reputedly hung, on their leader's instructions, from its boughs. Grigor recorded its girth at around 9m, considerably more than the Wymondham tree: it has not grown appreciably since. Its great size certainly suggests that it was alive at the time of Kett's rebellion, although there is no evidence that Kett or his rebels were ever active in this area.

There are fewer references to old named trees in the county in the twentieth century. Most of those mentioned by Grigor and other nineteenth-century writers seem to have been felled or have died. A number of large old trees do still have local names, although it is unclear how ancient these might be. Examples include the 'Salle Oak' in Salle, 'Big Fat Charley' in Gayton Thorpe, the 'Gospel Tree' in Wereham and the 'Squire's Oak' in Bradenham. New names continued to be coined – in Brundall, 'Margery Palmer's Oak' commemorates a woman who lived in a nearby cottage through much of the twentieth century – and trees continue to be planted to commemorate events or people, such as the oak planted in the churchyard at Kenninghall by the parish bellringers to commemorate the restoration of the bells in 1989. New traditions, linking old trees to local communities and their institutions, continue to be developed. Each year since 2001 a large or ancient tree has been chosen as South Norfolk District Council's 'Chairman's Tree'.

FIGURE 64. Kett's Oak in Ryston Park, on the edge of the Fenland of west Norfolk. This massive tree must be of medieval date, but there is no evidence that Kett or his rebels were active in this part of the county.

Grigor makes it clear that in Norfolk, as elsewhere, an interest in and affection for ancient trees is not new. Nor – in recent history at least – has it been something restricted to a wealthy or educated elite. Nevertheless, the sheer scale of the demand for wood and timber has ensured that, up until comparatively recent times, practical considerations generally took precedence over aesthetic and romantic ones, and relatively few trees were allowed to survive much beyond economic maturity. Standards were usually felled before they were a century old; pollards would have been taken down for firewood once their productivity declined in senescence. Local communities might, on occasion, retain ancient trees as boundary markers or as expressions of local identity, especially perhaps when located within villages, and Grigor's descriptions suggest that trees such as the great oaks which can be found today at Carbrooke and Old Buckenham, close to the centre of settlements, may be survivors from a once more extensive group. The great oak growing beside the Quaker meeting house at Tasburgh (erected in 1707) may be an example of something similar. But for the most part it was the wealthy elite who could afford to regard some trees primarily in non-economic terms, and who could thus retain them into old age. It is hardly surprising, therefore, although striking nevertheless, how many of the oldest and largest trees in Norfolk are to be found close to large mansions – in or on the edge of parks, lawns and pleasure grounds, or at the entrances to carriage drives. We would of course expect this of ornamentals and exotics such as lime, sweet chestnut, plane, beech and cedar. What is more striking is that the same is true of oak trees. No less than half of the oaks recorded in the county which have girths of 6m or more are found in such locations. In the vicinity of great houses old trees were evidently retained for their beauty or associations where in other contexts they would have been felled. Several other ancient oaks are found beside large farmhouses, perhaps retained by wealthy yeomen farmers for similar reasons. Numerous different forces have, over the centuries, militated against or encouraged the survival of old trees into senescence – water supply, soil fertility, exposure. But clearly, in an intensively settled county like Norfolk, human agency was the single most important influence on survival.

CHAPTER SIX

Orchards, Pine Rows and Willow Lines

Trees do not need to be very old, as we have already emphasised, to be classed as 'veterans' – much depends on the species. Nor do they have to be particularly large to be old. Not only do different species put on girth in very different ways, as we have seen, but particular forms of management can stunt the growth of trees, as for example when they are very closely planted and maintained by regular cutting, as in a hedge. In this chapter we examine briefly some of the classic Norfolk examples of trees which, while comparatively small in size, are nevertheless often of 'veteran' status, and constitute significant elements of the county's 'traditional' landscapes.

The Breckland pine rows

We will begin with the distinctive lines of Scots pines – pine rows, or 'deal rows' as they are sometimes known locally – which border many of the fields in Breckland, the acid, sandy district lying in the far south-west of the county, and extending across its boundary into Suffolk. Indeed, characteristic landscapes pay no heed to administrative boundaries, and in the discussion that follows we will examine Breckland as a whole and deal with the 'rows' in both counties. They vary greatly in appearance. Some are composed of trees that are tall and straight; others contain specimens which all exhibit a slight twist; while the most dramatic are made up of very twisted, picturesquely contorted pines (Figure 65). In some of the rows, moreover, the individual trees have girths of 3m or more but in others they are much smaller, sometimes only 0.5m in circumference. Some of the 'rows' are associated with low banks, but the majority are not.

The rows are a quintessential feature of the Breckland landscape, noted by many writers. To Ravensdale and Muir the 'tattered friezes of Scots pines ... are the hallmark of the landscape' (Ravensdale and Muir 1984, 204); while to Armstrong 'the isolated rows of Scots pine trees, standing out from the bracken of the remaining heathland, remains a typical feature of the East Anglian scene' (Armstrong 1985, 53). Most commentators agree that they were planted in the eighteenth or nineteenth centuries, when the open fields of the district were enclosed on a large scale and many of the local heaths reclaimed. They originally took the form of managed hedges, but they were later neglected and allowed to 'grow out', as the fortunes of landowners and farmers declined during the agricultural depressions of the late nineteenth and early twentieth centuries,

FIGURE 65. Variations in the Scots pine 'rows' of Breckland: (a) well-spaced trees with only a slight twist (Thetford); (b) slightly twisted and closely planted examples (Eriswell, Suffolk); (c) highly contorted examples growing beside a road (Gooderstone).

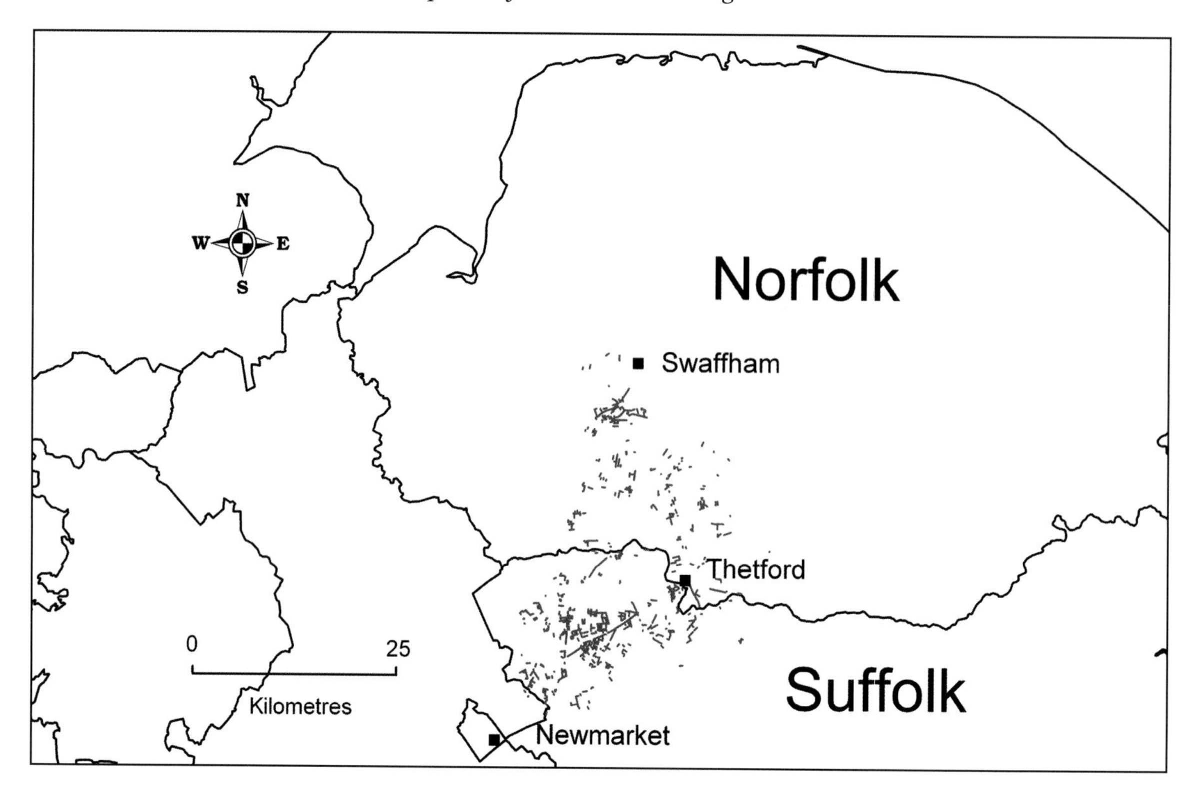

FIGURE 66. The distribution of the Breckland pine rows in Norfolk and Suffolk.

and especially when manpower was in short supply during the First World War. Today only a handful of examples are still recognisably hedges, mainly in Suffolk. Oliver Rackham neatly sums up the accepted history of these distinctive features of the landscape:

> A definite example of hedgerow trees arising by default are the rows of Scots pines, all much the same age and all gnarled at the base, which are the characteristic field boundaries of the Breckland. When the open fields and some of the commons of the Breckland were enclosed in the early nineteenth century, it was the fashion to make new hedges of pine … [but] even a few years' neglect causes pines to grow up irrevocably into trees. (Rackham 1986a, 223–4)

This 'late' origin of the pine rows has never, however, been unequivocally demonstrated, and some writers have favoured an earlier origin. In particular, Sussams and others have noted how in 1668 Thomas Wright described his use of 'Furre-hedges' to reduce the progress of the famous 'sand blow', or mobile dune, which engulfed much of the parish of Santon Downham in Norfolk (Sussams 1996, 105), thus implying that pine hedges were a traditional, long-lived feature of the local farming landscape. This in turn raises the possibility that the trees making up the surviving rows might be direct descendants of the Scots pines which – mixed with other native tree species – formed part of the natural vegetation of Breckland in remote prehistory. Conversely, Clarke in 1908 was confident that the rows had been planted on a large scale only 'since about 1840' (Clarke 1908, 563), thus implying a *later* date of origin than that proposed by writers such as Rackham.

The date of the pine rows has thus never been firmly established and a number of their characteristics have not yet been fully explained. It is not, for example, at all clear why their constituent trees exhibit such variations in growth patterns – why the pines in some are straight, and in others highly contorted. Another puzzle concerns the distribution of the 'rows'. Although often considered a quintessential feature of Breckland, they are by no means evenly distributed. Those familiar with the district will know the more noticeable 'hot spots': in and to the south of the parish of Cockley Cley in Norfolk (in the far north of Breckland), or in the area to the south-west of Elveden in Suffolk. Indeed, the location of the latter cluster, to either side of the A11 – one of the major gateways to East Anglia – may in part explain the prominence of the pine rows in popular images of Breckland.

The distribution of the pine rows

As with veteran trees of more conventional form, some light can be thrown on all these questions by combining the evidence of the available documentary sources with an examination of the distribution of the rows – that is, by studying their context. Mapping the rows immediately confirms the initial and subjective impression that there are major discontinuities in their distribution. Some of these can be explained by the presence of the Forestry Commission plantations, which were planted on a large scale on derelict arable and heathland in the district during the inter-war years. Any pine rows within these areas were presumably destroyed at the time or were cut down together with the rest of the trees in the plantations when these were first felled, before being replanted, sometimes in the second half of the twentieth century. Other *lacunae* are due to the presence of large military bases at Mildenhall and Lakenheath. But some of the gaps are evidently quite genuine: many parishes in Breckland which are characterised largely or entirely by agricultural land contain few or no pine rows, their fields being bounded by straight hawthorn hedges of the kind usually encountered in late-enclosed countryside. These places, moreover, often adjoin one or more parishes in which pine rows are common or frequent. One possible explanation for this pattern is that the pine hedges were only planted on particular types of soil and, indeed, a number of writers have suggested, or implied, that the Scots pine was favoured as a hedging plant in Breckland because it thrives in acid conditions, implying in turn that it would be more likely to be employed for enclosing the extremely acid land reclaimed from heaths than the more calcareous areas enclosed from the open fields. As Figure 67 indicates, however, there is little correlation between the distribution of the pine rows and that of particular soil types. While they are numerous on the leached, acid soils of the Newport 4, Methwold or Worlington Associations, they can also be found in some numbers on the more calcareous soils of the

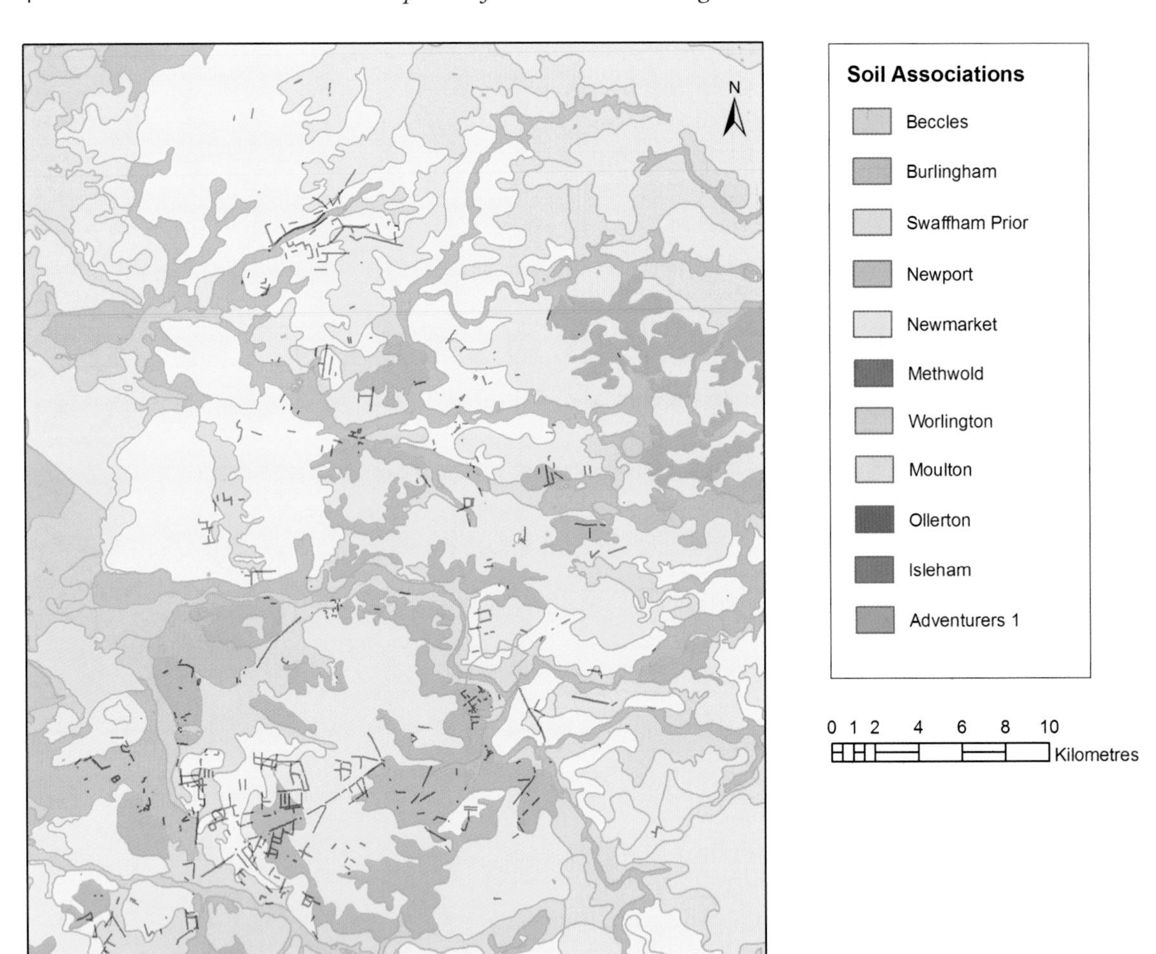

FIGURE 67. The distribution of the pine rows and principal soils types. Although many of the rows are found on the most acid soils of Breckland (those of the Newport Associations), they are also widespread on the more calcareous soils of the Swaffham and Newmarket Associations, and can even be found on the peat soils of the Isleham Association (the drained wetlands of the Fens lie immediately to the west of Breckland).

Newmarket 1 and 2 Associations – the better arable soils, farmed as open-field arable in the Middle Ages and in many places up until the nineteenth century. What is more surprising is that small but significant numbers also occur on the peat Fens immediately to the west of Breckland, especially in the Suffolk parish of Mildenhall. Evidently, the location of the pines was not determined in any simple or direct way by environmental factors, and the choice of pine as a hedging plant was one shared by some landowners living well beyond the boundaries of Breckland proper.

If comparing the distribution of the rows with environmental variables reveals few obvious patterns, comparison with aspects of administrative geography proves more revealing, for the main concentrations of pine rows are nested neatly within the boundaries of ecclesiastical parishes, as these existed in the early and middle decades of the nineteenth century: and in a more general sense there are signs of dramatic discontinuities at parish boundaries. Particularly striking in this respect is the concentration of rows, already noted, lying to the south-west of Elveden, for this is largely contained within the parishes of Elveden, Eriswell and Icklingham and, while extending into the northern parts of Mildenhall (the old township of Wangford) it stops abruptly at the boundaries of Lakenheath, Lackford, West Stow and Wordwell (Figure 68). In a similar way, the marked concentration around Cockley Cley in Norfolk – which accurate mapping reveals is actually centred more on the adjacent parish of Gooderstone – resolutely fails to extend into the adjoining parishes of Oxburgh, Foulden and Hilborough. The most obvious explanation for this relationship is that the presence or absence of the pine rows reflects the planting policy of particular landed estates at some point in the past, for these units were usually conterminous with parishes or groups of parishes. At first sight such an explanation seems convincing: the two main concentrations, around Cockley Cley and Elveden, could thus be associated with the large landed estates based on Cockley Cley Hall and Elveden Hall respectively. Closer inspection, however, soon reveals that this cannot be correct. Although the Elveden estate now includes much land in Eriswell and Icklingham, this was not the case in the early and middle years of the nineteenth century. Icklingham was then owned, almost in its entirety, by the Gwilt family, while Eriswell was in divided

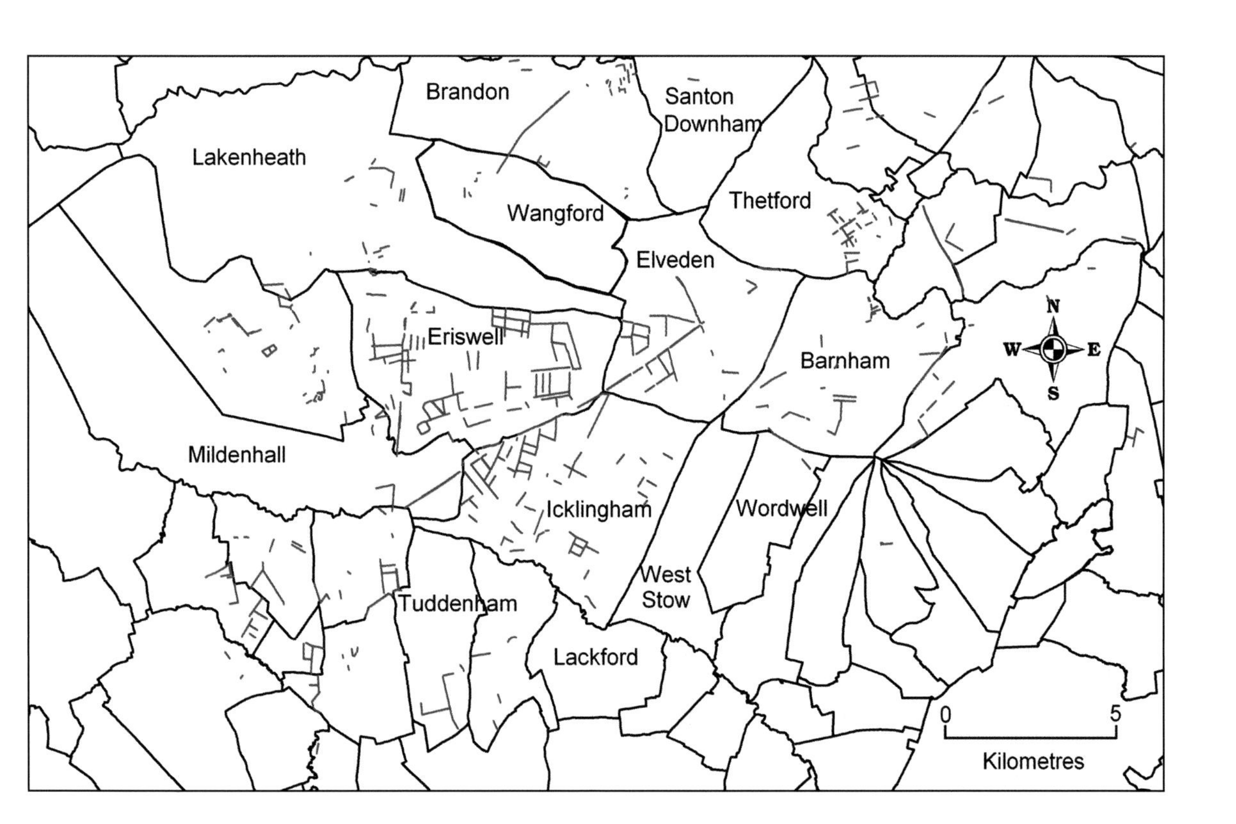

FIGURE 68. Pine rows and nineteenth-century parish boundaries in the area to the south and west of Thetford.

ownership, but with the Company for the Propagation of the Gospel in New England, no less, as the main owner (Postgate 1960).

The age of the pine rows

An alternative explanation would link the distribution of pine rows to the chronology of enclosure – that is, the replacement of a landscape of arable open fields, periodically cultivated outfield 'brecks' and heaths with one of enclosed fields owned and occupied in severalty: for in general terms each parish, in East Anglia as elsewhere, has its own enclosure history, distinct and often very different from that of its immediate neighbours, so that the age of field boundaries can change dramatically on either side of a parish boundary. Unfortunately, the enclosure history of Breckland is particularly complicated. Because many settlements, especially in the centre of this arid region, experienced severe contraction or desertion in the late Middle Ages, they often became the sole property of particular proprietors and were thus *technically* enclosed – in the sense that they had a single freehold owner and no commoners, with common rights – long before the eighteenth century. However, the date at which (and extent to which) they were enclosed in *physical* terms displays much variation. Some, such as Buckenham Tofts, shown on an estate map of 1700 (NRO Petre Mss Box 8), were largely divided into fields at an early date, but others often remained physically unenclosed into the nineteenth century, their arable land remaining as open fields, with intermixed strips held by tenants, while their heaths continued to lie open and unreclaimed. Such landscapes were often transformed during the early nineteenth century, when high grain prices and a fashionable concern for 'improvement' saw the removal of common arable and large-scale (if often short-lived) reclamation of heathland.

Other parishes, particularly towards the perimeter of Breckland, were enclosed by parliamentary acts in the late eighteenth and nineteenth centuries, with the actual award – the fixing of the new pattern of boundaries and the division of the land between the various proprietors and common-right holders – usually coming between three and eight years later, after the areas in question had been surveyed and the legal claims of those seeking an allotment of enclosed land fully investigated. By the terms of each award, the allotments given to landowners in lieu of their holdings in the open fields and their lost common rights generally had to be fenced or hedged within twelve months. This does not, however, mean that every boundary in a parish dates to the year of, or immediately following that of, the award. This is because the allotments of land made to the major proprietors in a parish were often very big, and were only later subdivided into more manageable fields. Indeed, whenever or however enclosure was achieved, further modifications to the field pattern were often made – hedges added, or taken away – throughout the nineteenth century.

It is against this rather complex background of local enclosure history that we need to examine the distribution of the pine rows. On Figure 69 the Breckland

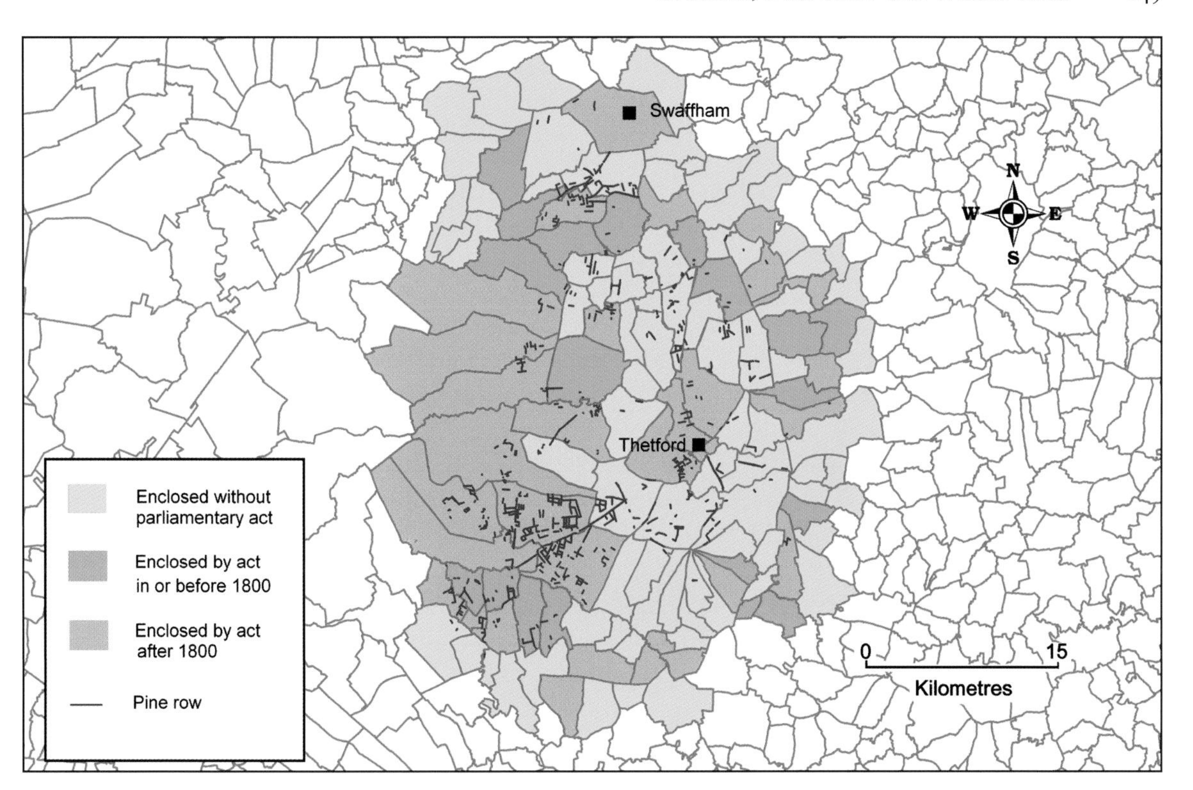

FIGURE 69. The distribution of the pine rows and the enclosure history of Breckland parishes. The rows are strongly associated with parishes enclosed by parliamentary act after 1800.

parishes are divided into three categories: those enclosed by parliamentary awards dating to the period before 1800; those with parliamentary enclosures after 1800; and those enclosed by non-parliamentary means. It is immediately apparent that pine rows are rare in the first category, occur to very varying extents in the third, but are very closely correlated with the second: a strong indication that they were largely, if not entirely, planted in the period after 1800. We can, however, refine this chronology further. The two parishes which contain the largest numbers of rows, Eriswell and Icklingham, were enclosed in 1818 and 1816 respectively. Other parishes with significant concentrations were enclosed around the same time: Freckenham in 1820, Feltwell in 1815 and Hockwold in 1818. Indeed, only three parishes with significant concentrations of pine rows were enclosed by earlier awards: Thetford in 1806, Gooderstone in 1805 and Methwold in 1807.[1] This strongly suggests that the peak period for planting was in the years between 1815 and 1820, but leaves unresolved how much earlier the practice actually began.

Closer examination of the Gooderstone enclosure map (NRO C/Sca 2/132) provides further useful information. It is immediately apparent that relatively few of the pine rows correspond with the boundaries of the actual allotments made by the enclosure award. Instead, they mainly form subdivisions of allotments. The exceptions, moreover, in several cases run along only part of a boundary between adjacent allotments, perhaps indicating that they represent later replacements for hawthorn hedges which had failed to thrive on this light,

[1] Mundford was enclosed by an act of 1806 but the date of the award is uncertain.

acid land. At the very least, the pattern implies that the use of pine hedges was only just beginning when the enclosure took place in 1805, and it is noteworthy that with few exceptions the 'rows' are found within, or on the boundaries of, the land of only one of the several proprietors to benefit from the enclosure of the parish, the dowager duchess of Essex. Her local agent, perhaps, was an individual fully abreast of the latest agricultural fashions.

As already noted, parishes enclosed by non-parliamentary means often have poorly documented enclosure histories, but some light can be thrown on the antiquity of their field boundaries by using larger-scale maps: most notably, William Faden's map of the county of Norfolk, published in 1797; Hodskinson's Suffolk map of 1783; and, in particular, the draft 2-inch to the mile drawings prepared by the Ordnance Survey at various times between 1810 and 1820 but never printed and published (until revised at a scale of 1 inch to the mile several decades later). None of these maps shows boundaries with any accuracy, although the Ordnance Survey drawings shows field patterns in a partial and schematic way. All, however, carefully distinguish between roads enclosed by hedges and those which were unenclosed, the bounds of the latter being depicted with a dashed line. Faden's map and the Ordnance Survey draft also show areas of surviving heathland. On Figure 69, two parishes enclosed by non-parliamentary means stand out as having particularly dense concentrations of pine rows: Elveden in Suffolk and Cockley Cley in Norfolk. In Cockley Cley, most of the pine rows occur on roadsides, although a substantial minority are field boundaries. Almost all the roads in the parish, and all the roads on which the rows appear, are shown as unenclosed on Faden's map and also on the Ordnance Survey 2-inch draft. The fact that the roads were unenclosed indicates that much if not all of the land lying between them also lay open, as open fields or common grazing; and some of the pine rows clearly lie within areas explicitly shown on both maps as open heath. In Elveden the situation is similar. Once again the pine rows occur on the sides of roads shown as unenclosed both on Hodskinson's map of 1783 and on the Ordnance Survey draft drawings. Here, however, the maps also indicate more clearly that most if not all of the areas in which the other rows were located were occupied by open heaths.

Less pronounced clusters of pine rows appear in Didlington, West and East Wretham and Snarehill, and here the pattern is repeated. Pine rows occur on roadsides shown as unenclosed on the various large-scale maps from the late eighteenth and early nineteenth centuries, and within areas clearly depicted as open heathland. Only a few examples occur in other locations, and there is little reason to doubt that these, too, still lay unenclosed, as open fields, at the time. In short, the cartographic evidence from these less well-documented parishes, such as it is, supports that from places enclosed by parliamentary act in suggesting strongly that few if any of the pine rows originated before 1800.

The only piece of evidence that has been produced to suggest an earlier date for pine hedges in the region, and which implies that they might be a 'traditional' form of enclosure, comes from the late seventeenth century. As already noted, Kate

Sussams and others have noted how as early as 1668 Thomas Wright described his use of 'Furre-hedges' to reduce the progress of the famous 'sand blow' that engulfed much of the parish of Santon Downham (Sussams 1996, 105). But it is difficult to believe that a landowner would improvise a barrier to a moving sand dune by *planting* a hedge of any kind – the sands were moving quickly, and would not wait while the plants matured into a significant obstacle. Almost certainly, Wright was describing the creation of 'dead hedges' of staked gorse, a plant locally known as furze. Gorse was widely used in the area to make, or gap up, fences or hedges. John Norden, who had long experience of East Anglia, described in his *Surveyor's Dialogue* of 1618 how it was used 'to brew withall and bake, and to stoppe a little gap in a hedge' (Norden 1618, 234). What is particularly striking in this context is that pine hedges receive no mention in Nathaniel Kent's *General View of the Agriculture of Norfolk* of 1796, in Arthur Young's 1804 book of the same name, or in the same writer's *General View of the Agriculture of Suffolk* of 1797. This is noteworthy, as Young in particular was usually keen to describe novel or unusual farming practices. In contrast, Richard Bacon, in his *Report on the Agriculture of Norfolk* of 1844, refers to 'fences of Scotch fir' as if they were a well-established part of the Breckland scene (Bacon 1844, 392).

The evidence thus suggests that pine hedges were first planted in the region some time after 1800, and that the peak period of planting was between 1815 and 1820. The fashion may have continued on some scale into the 1820s, to judge from the only surviving description of the planting of the rows so far discovered. In 1829 the Suffolk traveller David Davey described how:

> Within two miles of Brandon I observed a mode, to me at least new, of raising a good fence in a very bad soil; a bank is thrown up, about four or five feet high, and of a considerable thickness at the bottom; upon the top of this is planted a row of Scotch firs, as thick almost as they can stand; these seem to make rapid progress in this soil and branching out towards the sides, immediately from the ground, and have the additional very strong recommendation of affording the best shelter from storms to the sheep and cattle which are fed, or rather starved upon the land. (Blatchley 1982, 136)

There are also signs that hedges continued to be established, on a sporadic basis, through the nineteenth and into the twentieth century. Those bordering the A11 to the south-west of the Elveden Monument, for example, do not appear to have existed in 1880, as the road is shown on the Ordnance Survey first edition 6-inch map of that date as unenclosed. For the most part, however, the rows are clearly a phenomenon of the years around 1815–20.

The abandonment of the pine rows

As we noted earlier, there appears to be a universal consensus that the rows were all originally established as hedges, the management of which was gradually neglected and abandoned in the period after *c.*1880. W. G. Clarke, writing in

1908, implies that most of the pine hedges were still being regularly managed at that time: 'These hedges are made of ordinary trees kept stunted by constant trimming.' The only exceptions he seems to note are 'many of the lines of fir trees now bordering plantations' which 'were originally hedges, but have ceased to be trimmed' (Clarke 1908, 563–4). As late as 1925 the same writer implied that many were still being regularly cut, noting how pines were 'still the characteristic tree of the district, planted either in rows known as "belts", or artificially dwarfed as hedges' (Clarke 1925, 17). A photograph by Clarke in the Norfolk Record Office, dated 1923, shows an example of a managed hedge in Ickburgh. By the 1930s, however, Butcher implies that many had been abandoned, describing how 'the rows of trees are in all stages of development and degeneration: there are some very fine hedges along the Barton Mills–Elveden road and some very neglected ones around Cavenham and Icklingham which no longer fulfil their primary purpose' (Butcher 1941, 353).

There is, however, abundant evidence that a significant proportion of the rows had in fact been allowed to grow out long before this. A number of photographs showing storm damage on the Didlington estate in Norfolk, taken around 1900, show what appear to be tumbled lines of pine trees (NRO MC 84/133, PH7). More importantly, a significant number of the rows are shown as lines of conifers on the Ordnance Survey first edition 6-inch maps from the 1880s. As already noted (above, p. 27), the surveyors were instructed only to show mature trees, well beyond the sapling stage, thus indicating that the management of the rows in question must have been abandoned many years earlier (Figure 70). In this context, attention should again be drawn to the marked variation in the growth pattern of the trees in different rows, ranging from almost upright, with only a slight twist, to excessively contorted (see Figure 66). Without exception, it is those rows made up of trees towards the 'straight' end of the range which are shown as lines of individual conifers on the Ordnance Survey first edition 6-inch maps. Those composed of the more contorted examples are not so shown, indicating that they were still being managed as hedges at the time. This raises the possibility that some of the pine rows may always have comprised lines of separate, well-spaced trees, rather than managed hedges, but fieldwork suggests otherwise. In many cases, lines composed of 'straight' trees include, interspersed with the full-grown specimens, stunted examples and dead stumps, sometimes so closely spaced that they actually touch. In the example shown in Figure 66(b), for instance, in Eriswell, the live pines vary in size from 0.4m to 1.6m in circumference at waist height, with dead examples and stumps ranging from 0.3m to 0.6m. Although displaying only a slight twist, and already depicted as a line of trees on the Ordnance Survey 6-inch map, this row was clearly planted as a hedge but, like the others in the immediate vicinity and many others in the area, was clearly not managed as one for very long. Some of the pine rows boast trees with rather larger circumferences, in rare cases reaching as much as 3.7m. It is possible that, in these cases, some of the intervening trees were actively removed at an early stage, providing extra light and space for the remainder. In

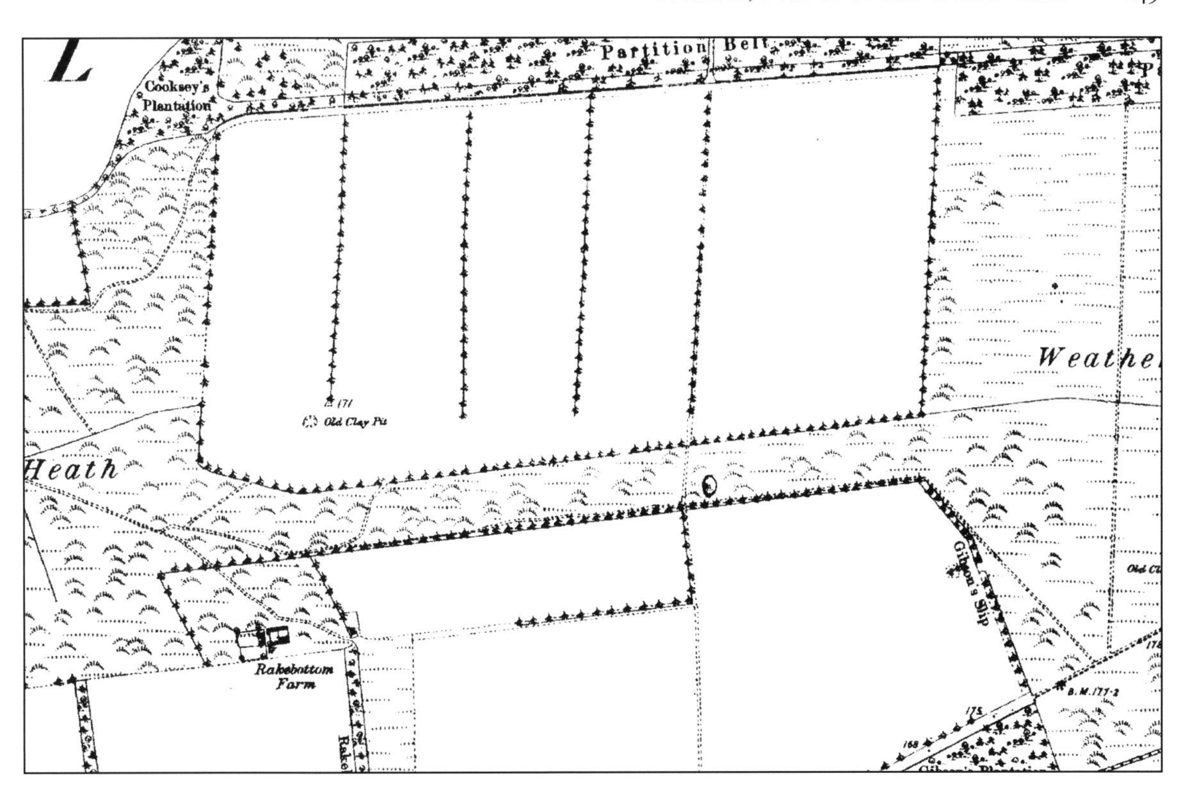

FIGURE 70. Extract from the Ordnance Survey first edition 6-inch map of 1891 showing part of Eriswell in Suffolk. Many of the pine rows had already become lines of mature trees.

most cases, however, it appears that the rows have 'thinned' naturally, through the success of some trees and the gradual failure of neighbours.

The significance of the pine rows

The pine rows of Breckland are interesting, from the perspective of the landscape historian, for a number of reasons. Although they are now a much-loved, characteristic, even 'traditional' feature of the local landscape, they appear to be the result of a relatively recent and probably short-lived fad or fashion. Somebody – presumably a local nurseryman and possibly Mr Griffen, who ran a business based in Mundford, more or less in the centre of the distribution of the 'rows' – evidently persuaded many local landowners in the early nineteenth century that pines would make an ideal hedging plant on these dry and difficult soils. But the fact that a high proportion of the closely planted lines of pines were managed as hedges for only a short period of time, if at all, suggests that many people realised almost immediately that Scots pines do not make good hedges, in an agricultural context at least. Pine hedges are more difficult to plash or lay than ones of hawthorn, easily bare out at the base, and need regular trimming to prevent them developing into lines of separate trees. Once they have made this transition, moreover, it is impossible to return them to the desired state. The fashion may have lasted no more than a dozen years, and, although some further pine hedges were planted later in the nineteenth century, the practice was evidently rare, and perhaps largely aesthetic in character, motivated by a desire on the part of some landowners to match the appearance of existing boundaries in a particular locality.

In terms of the long history of the East Anglian landscape the pine rows are thus relatively recent features. But compared with most Scots pines in East Anglia the individual trees of which they are composed are old: around 200 years old, and thus more than twice the age of the oldest pines within the Breckland plantations established by the Forestry Commission. Whether the smaller examples, in particular, should be regarded as of 'veteran' status may be debated, but the 'rows' as a whole unquestionably play an important part in sustaining local biodiversity. The pines form corridors for wildlife through what is often an intensively arable landscape, form important feeding territories for hawfinches and crossbills and appear to be favoured by bats (pipistrelle and barbastelle especially) and perhaps barn owls (Rothera 1997). They are associated with, and help preserve from arable encroachment, strips of tussocky, ungrazed grass, often of such a width that their central areas, at least, are only marginally affected by chemical sprays, and which are frequently crossed by tracks or disturbed by rabbits, thus perpetuating the ruderal conditions in which

FIGURE 71. A highly contorted row at Tuddenham in Suffolk. With their abundance of dead wood and their associated strips of unploughed, but often much disturbed, grass, the pine rows make an important contribution towards maintaining the distinctive flora and fauna of Breckland.

the most characteristic of the Breckland flora thrive, having developed not in conditions of permanent heathland but rather in that of the shifting, periodic cultivation which characterised the arable 'brecks' of the district (Dolman 1994). The grass strips are also rich in insects, including such nationally rare species as the ground beetle *Ophonus laticollis*, and favoured by skylarks, grey partridges and other farmland birds, as well as by hares. The pines themselves also provide habitats for a wide range of insects. The bark of the larger trees forms layered plates separated by deep fissures, and dead wood is particularly abundant given the fact that, as already emphasised, the rows frequently contain the stumps and broken trunks of failed trees, squeezed out by competitors; and, in the case of the more twisted and contorted examples, feature an abundance of fallen and broken branches (Figure 71). In environmental terms, the pine rows are a particularly happy accident of history.

Willow rows

Similar in many ways to the pine 'rows' of Breckland, although less well known, are the lines of close-set pollarded willows which can be found beside many of the principal roads in the far east of the county, in the low-lying drained marshlands in the lower reaches of the Broadland rivers – on the marshes of Halvergate and in the adjoining areas of reclaimed wetland. When maintained in the traditional way – that is, regularly pollarded at a height of around 1.5–2.0m – these form lines of continuous vegetation not dissimilar in appearance to an outgrown hedges (Figure 72). Inspection reveals that the willows were originally planted at intervals of around 4m on the edge of the verge and immediately above the ditch flanking the road. In some places – as on the branch road to Halvergate

FIGURE 72. Halvergate, east Norfolk: this nineteenth-century road, leading south from the Acle Straight (the modern A47) towards Halvergate village, is flanked by well-preserved examples of the willow lines characteristic of the district.

– the trees are well preserved on both sides of the road. Elsewhere, as on the nearby 'Acle Straight' (the A47 to the west of Yarmouth), their survival is patchy, many of the trees having been removed in comparatively recent times (the RAF aerial photographs of 1946 show them as being much more continuous).

The willow rows appear to be very much a part of the 'traditional' landscape, and they are, albeit perhaps to a lesser extent than the Breckland 'rows', a characteristic feature of the local countryside. But in reality, like the 'rows', they have a fairly recent origin. Some early maps of the area show lines of pollards: thus a map made in 1749 of an area of marsh to the west of Reedham church shows a number growing beside drainage dykes (NRO MC 27/1, 501X4), while another, surveyed in 1825, shows several probable pollard willows growing beside tracks and field boundaries around Oulton Dyke near Lowestoft (Barnes and Skipper 1995). But in general early maps suggest that the marshes were largely or even completely devoid of trees. Nineteenth-century paintings of the marshes similarly attest their open, treeless quality. Once again, careful inspection of the context of the trees provides immediate clues to their origins. With few if any exceptions, the willow rows are all associated with roads and tracks which were either newly created or extensively improved in the nineteenth century. Local tradition is unanimous regarding their purpose: the willows were planted to 'hold the shoulder of the road' – that is, to prevent the road from subsiding into the flanking ditches. The trees were regularly pollarded in order to keep them low and so prevent them being brought down by the wind – and thus wrecking the surface of the road. Pollarding also allowed the trees to be narrowly spaced, and thus maximised the amount of root structure available to bind the ditch banks (Barnes and Skipper 1995, 200). It is possible that the technique was employed in the area before the nineteenth century, but unlikely, given the evidence of early maps already noted. The willows are an important and characteristic feature of the Broadland scene, and a striking feature in the more extensive tracts of level marshland, as on Halvergate. But they were planted with a straightforward, practical purpose, and not very long ago.

Orchards

Old orchards, and the traditional varieties of fruit trees which they contain, have been the focus of much interest and concern from conservationists and others over the last few decades (Keech 2000; Morgan and Richards 2002; Sanders 2010). Organisations such as the East of England Apples and Orchards Project and Common Ground have done much to preserve old orchards and to establish new ones. Maintaining a wide range of fruit varieties against the increasing commercial concentration on a few is seen by many as important for promoting biodiversity. But, in addition, old orchards are themselves important habitats. Traditionally, fruit trees were managed as standards or half-standards, and cropped for many decades. Apples and pears reach 'veteran' status in a relatively short time – less than a century in some cases – and then provide the usual range of habitats in the form of crevices, hollows, rotten wood and the like. In addition, the areas beneath the trees usually constitute undisturbed and unimproved grassland which, managed by cutting or grazing, is floristically rich. Traditional orchards represent, in effect, a variety of wood-pasture, and can provide vital habitats – not least for farmland birds – within what are otherwise often intensively farmed landscapes. It is hardly surprising, then, that orchards were, in 2007, included in the UK Biodiversity Action Plan, having already been included in those drawn up by a number of individual counties.

There are always dangers in romanticising any element of the 'traditional' landscape, and orchards are no exception. Most of the larger examples in East Anglia are in fact of no very great antiquity. The number of commercial orchards grew steadily through the middle decades of the nineteenth century, as improvements in communications – principally the construction of a national rail network – allowed perishable commodities like fruit and vegetables to be transported to the expanding urban and industrial areas with relative ease. In the period after *c.*1880 growth accelerated, as the severe agricultural depressions of the late nineteenth and early twentieth centuries encouraged many farmer and landowners to diversify production, replacing a dependence on wheat and barley with the cultivation of a wider range of crops. The area devoted to orchards in Norfolk thus rose from 6345 acres in 1914 to 8414 in 1926 and 10,089 in 1936 (Mosby 1938, 183). They came to be particularly important in two main areas: the north-east of the county, where expansion seems to have begun in the 1890s, spearheaded in part by major landowners such as the Cubitts of Honing and the Petres of Westwick; and the silt Fens, where the industry was already expanding fast by the 1840s (Wade Martins and Williamson 2008, 49–51). This area formed part of a wider fruit-growing district which embraced the adjacent parts of Cambridgeshire and Lincolnshire and was centred on Wisbech:

> This area extends from Upwell in the south to Terrington in the north, being about a mile in width in the south, it widens rapidly in the neighbourhood of Wisbech and exceeds 6 miles in the north at Terrington. The older orchards are to be found near Wisbech and the more recent extensions in the north, but so far the orchard area does not extend more than half a mile north of the Lynn–Sutton Bridge road. East of Terrington the continuity of the orchards is broken by grass and arable land, but there is a big concentration of orchards on both sides of the Great Ouse in the Wiggenhalls. (Mosby 1938, 183)

By the inter-war years the Fen orchards were mainly producing cooking apples and plums, which were sent to the north of England or used in local jam factories. The orchards in the east of Norfolk were more widely scattered and grew dessert apples, such as the Worcestershire Pearmain and Cox's Orange Pippin. Some orchards supplied apples to the Gaymers' cider factory, which had been set up beside the railway line at Attleborough in 1896 (Mosby 1938, 184).

Many of these large commercial orchards still remain, especially in the Fens, in spite of a marked decline in the industry over the last three decades. Most are regularly and fairly densely planted, either with half-standard trees or, more usually, with dwarf trees in 'spindle' form, a method of management which became popular with growers in the 1970s. Either way, the trees tend to be of no great age as they are usually removed when they reach senescence and their productivity declines. Many commercial orchards, moreover, have a floristically poor sward which is managed by chemical sprays. They have a certain value in maintaining biodiversity, especially in the Fens – now an intensively arable landscape largely devoid of permanent grass, trees or hedges; and some of the longer-established half-standard orchards are developing into mature orchard habitats. But the larger, commercial and more recent orchards cannot in general compete, in terms of biodiversity, with traditional ones (Figure 73).

It is the latter – small, and mainly geared towards domestic consumption – which are our main concern here. They have been in decline for more than a century – the Land Utilisation Survey of 1938 contrasted them with the new commercial concerns, describing them as 'small, unimportant, and usually old and neglected' (Mosby 1938, 183). But at the end of the nineteenth century they were a commonplace feature of the landscape and a vital part of the domestic economy. The Ordnance Survey first edition 6-inch maps of the 1880s show vast numbers, while smaller collections of fruit trees – too small to be mapped – were associated with many cottages and most smaller farms. Domestic orchards have a long history in Norfolk. As early as 1089 the foundation charter for Castle Acre Priory refers to the gift of two orchards, while numerous descriptions of properties over the following centuries refer to them. An extent for Langley Abbey dating to 1289, for example, describes 'the mansion house containing 11 acres, with the out ditches, mote etc: apples in the orchard, valued at 6s 8d' (Blomefield 1805, vol. 8, 356–77; Dallas 2010b, 4). Much of our evidence, as always, is biased towards the larger residences – manors, monasteries and castles – but the fruit trees growing on smaller properties are also sporadically mentioned in medieval sources, as, for example, in grants of farms at Hockering in 1386 (NRO EVL 396/2, 416X4) and Great Melton in 1391 (NRO EVL 189, 455X1). But it is only as documents providing detailed descriptions of property become more abundant from the sixteenth century, and include maps and plans as well as written accounts, that we can learn much about their management and contents.

Such sources reveal that although apples always predominated over other fruit (cherries, plums, nectarines, peaches, gages, cherries, walnuts and filberts) in the county's orchards they did so to a lesser extent in the period before *c.*1800, when pears in particular were also a prominent feature. An order for fruit trees at Ryston Hall in 1672 thus mentions twenty-four apples, but also eighteen pears, while a lease for Quebec Hall in East Dereham from 1784 records twenty-two apples and eleven pears growing in the grounds, together with fifteen cherries, twelve plums and three filberts (NRO MF 219/11; NRO BUL/16/230. Indeed, pears made up around 40 per cent of the fruit trees mentioned in thirteen detailed sources in the Norfolk Record Office dating from the period 1600–1790, but only 13 per cent of those described in sources from 1790–1900 (Dallas 2010b, 11).

FIGURE 73. Traditionally managed orchards, with veteran trees and grassland untreated by chemicals, are rich in wildlife.

The orchards and fruit grounds of country houses in the seventeenth, eighteenth and nineteenth centuries boasted a phenomenal range of varieties of fruit. At North Elmham, for example, the kitchen garden newly planted in 1765 by Robert Milles contained thirteen different varieties of cherry, six of nectarine, five of apricot, twelve of peach, nineteen of pear, fourteen of plum and two of apple, as well as a medlar. Ninety-eight individual trees were planted in the garden and a further fifty-six on the outside walls and in adjoining areas (Roberts 1937; Williamson 1998, 164). The kitchen garden laid out at Shotesham in the late 1780s contained four varieties of nectarines, four of apricot, eight of plum, nine of cherry and fifteen of peach (NRO FEL 1115, L5) (in both these cases, apples and pears were concentrated elsewhere, in an orchard). The bills for fruit trees purchased for Heydon between 1792 and 1801 mention seventeen different varieties of peach alone (NRO BUL 4/140, 610X5; BUL 11/89). But even yeomen farmers and local clergy could amass significant collections. Thomas Ripingall of Langham, for example, listed in the early nineteenth century the four varieties of nectarine, four of peaches, two of plum and two of cherry on his farm (NRO MC 120/45, 593X4). In general, peaches, nectarines and apricots were a more important feature of the grounds of large mansions than of small farms, largely because successful cultivation required the kind of shelter that could only really be provided by expensive walled enclosures. Plums were proportionately more important in the orchards and gardens of farmers, where they were often planted around their edges.

Orchards had other uses beside the production of fruit. Those at large mansions were often partly ornamental in character, especially in the sixteenth and seventeenth centuries. At Stow Bardolph in 1712 apples, pears and plums were all planted in 'the quarters of the wilderness'. Farm orchards were regularly cut for hay (there are, for example, several payments for orchard hay in the detailed tithe accounts for Shotesham which survive from the seventeenth and eighteenth centuries) (NRO FEL 476–480); and they were also grazed – by sheep and geese, but probably not by cattle or horses. It is often said that pigs were kept in orchards but if so they were presumably housed in styes. If allowed to roam freely, browsing on windfalls, they would soon have damaged the roots of the trees.

Traditional varieties of fruit

Not only are many of Norfolk's surviving orchards of relatively recent origin: so too are many of the county's 'traditional' fruit varieties. Of the fifty-three varieties of apple identified as important by the East of England Apples and Orchards Project, for example, no fewer than twenty-nine first appeared in the period between 1870 and 1950. Some of these were the result of experimentation at (or natural selection arising in) the orchards and kitchen gardens of country houses, but many were created by the large commercial orchards – such as the Emneth Early and Lynn's Pippin, both developed by the Lynn family of Emneth in 1899 and 1942 respectively. This said, a significant proportion of the distinctive fruit varieties still to be found in the county's orchards have a respectable antiquity and documentary sources from earlier centuries name numerous others, although it is apparent that different names were often given in different documents to what are probably the same varieties of fruit.

Specific varieties of apple are mentioned even in medieval sources. Pearmain apples, for example, are referred to in documents relating to property in Runton and Rackheath in 1204 (Blomefield 1805, vol. 11, 241–6; vol. 10, 446–51), while the words 'Pearmain', 'Codling' and 'Costard' regularly appear in Norfolk probate records throughout the fourteenth and fifteenth centuries. But it is, once again, only from the early seventeenth century that our records become sufficiently detailed to reveal a more extensive list (Dallas 2010b). In all, no fewer than 253 varieties of apple and 140 of pear are known from seventeenth-, eighteenth- and nineteenth-century sources from the county (in the lists of fruit growing on or purchased for particular properties, or in the catalogues of the various nursery companies, most notably that produced by George Lindley of Norwich at the end of the eighteenth century). Some of these were widely grown in England. Indeed, even the famous Norfolk Beefing may be less indigenous than we often assume, for the earliest forms of the name – *Biefen* and *Beaufin* – have a continental appearance, and both Yorkshire and Lincolnshire apparently grew identical apples; while the Norfolk Pippin seems the same as the Hanging Pearmain grown in Herefordshire. Others, however, including Winter Majetin, Five Crowned Pippin or Pine Apple Russet (Hogg 1851), clearly originated in and were a particular speciality of the county. The documents also refer to a number of varieties which are now lost, including the Thwaite, the Free Thorpe, and the Halvergate, which are all mentioned in a planting list prepared by Mary Birkhead for her daughter's orchard in Thwaite in 1734 (NRO BRA 926/121/2); and the Oxnead Permain, probably first developed on the Earl of Yarmouth's Oxnead estate in the late seventeenth or early eighteenth century.

The large number of different apple and pear varieties recorded in planting lists and other sources for the county in part reflects a desire on the part of owners to provide a wide range of different tastes at a time when the range of confectionaries now widely available was then unknown. Different apples and pears were used in different ways. Some, such as Acklam's Russet, Early Nonpareil and Stone Pippin, were eaten raw; others, including Baxter's Pearmain, Catshead, Dr Harvey and Kitchen Rennet, were mainly used for baking or boiling; while others, such as Marmalade Pippin, Royal Wilding, and Summer Red-Streak, were particularly good for cider-making. In addition, and perhaps more importantly, landowners and farmers wished to ensure that fruit was available over a long period: thus, while some apples and pears needed to be eaten soon after they were picked, others were good 'keepers' and might even be left on the tree well into the winter. The aim was to produce both a small quantity of dessert and culinary fruit for consumption in the late summer and autumn and also a range of varieties which could be preserved or stored into the following spring. Not all apples, pears and other fruit were necessarily consumed by the household that grew them, however, for they were often used as gifts to friends, neighbours or the local poor. Fruit could even be used almost as a form of currency, to pay rent or mortgages: a memorandum of 1701 records the lease of a parcel of land in Downham Market in west Norfolk 'for the payment of 3lbs of potatoes and the fruit of three apple trees each year to Thomas Buckingham and his wife for their lives' (NRO SF431/19, 308X5).

The layout and distribution of traditional orchards

Orchards, at least before the development of large commercial concerns in the nineteenth century, were normally located close to the house. From the sixteenth century onwards writers provided detailed advice on how they should be arranged (Markham 1613; Lawson 1618). The layout of one early Norfolk example is recorded in some detail in the description penned by Mary Birkhead of her daughter's orchard at Thwaite in 1734 (NRO BRA 926/121/2). This covered around an acre and was surrounded by a fence, the inside of which was planted with plums, quinces, barberries, nuts and filberts. There was a row of walnut trees along one side of the orchard (probably the north), and filbert bushes were placed at intervals between the trees, but the orchard was principally composed of apples (thirty-six trees, each a different variety) and pears (six trees, six varieties), which were arranged in a neat grid of nine rows by six, with the trees spaced thirty-six feet apart in one direction and twenty-six feet apart along the diagonal lines. Mary provided a rough plan showing this arrangement. Another orchard described in her memorandum book was larger, with eleven rows and a total of 152 trees. Once again, apples were the most numerous trees, but the orchard included pears, filberts, walnuts, cherries and plums.

Recent work by Patsy Dallas has revealed important spatial variations in the numbers and character of domestic orchards in Norfolk by the later nineteenth century (Dallas 2010b). As already noted, by *c.*1880 large commercial orchards had already begun to develop in the Fens: indeed, some had appeared as early as the 1840s, with, for example, thirty-eight hectares of orchard being recorded in the tithe award for Walsoken in 1843 (NRO DE TA 33). Interestingly, the locations of the Fen orchards do not appear to have been very stable: successive maps through the nineteenth and early twentieth centuries tend to show old areas reverting to arable while neighbouring arable fields are converted into orchards. Within a sample area of 225 square kilometres there were, when the Ordnance Survey 6-inch maps were surveyed in the 1880s, 131 orchards (0.58 per square kilometre), of which 66 covered between half and two hectares, 42 less than half a hectare, with the rest being substantial concerns, some examples extending over as much as thirteen hectares. The situation on the claylands in the south-east of the county was very different. Here there were far more orchards – as many as 1.4 per square kilometre – but the vast majority were small and domestic in character, with around 40 per cent covering less than half a hectare and almost all the others being less than two hectares. Different again was the situation in the north-west of the county, and in the north, where there were much smaller numbers – between 0.1 and 0.2 per square kilometre within the two sample areas of 225 square kilometres, none larger than two hectares. In Breckland there were slightly more, as many as 0.3 per square kilometre, but still significantly fewer than on the claylands of the south-east. The explanation for these patterns is not entirely obvious. The suitability of the rich silt soils and comparative proximity to markets in the Midlands and north presumably explains the early development of commercial production in the Fens – something which, as we have noted, was to increase markedly in the years after *c.*1880. The large number in the south-east of the county may also to some extent reflect the character of the local soils, but is perhaps mainly to be explained in terms of landholding patterns. This, as we have repeatedly emphasised, was an area in which there were very large numbers of small owner-occupiers, for whom the planting of an orchard – a medium- if not long-term investment – made obvious sense. In the north-west and in Breckland, and to some extent in the north of the county, there were fewer freeholders, more large estates and fewer, larger, tenanted farms. Gentleman farmers, holding on relatively short leases, were perhaps less inclined to invest in the creation of orchards than were their yeomen brethren on the southern claylands.

The recent history of Norfolk orchards

Norfolk orchards, both commercial and domestic, have not fared well in recent decades. Only around 3 per cent of those shown on the Ordnance Survey first edition 6-inch maps of the late nineteenth century remain in production, with a further *c.*10 per cent surviving in partial or relict form. Much of this decline has happened in recent years: in *c.*1950 there was over 4000 hectares of orchard in the county; today there is less than 500 hectares, mainly in the far west of the county, on the silt fens (East of England Apples and Orchards Project 2006). The greatest losses, in terms of 'traditional' orchards, have been to housing, as village 'envelopes' have been progressively infilled; large numbers have also been converted to gardens and paddocks by owners more interested in flowers and horses than in fruit. But major losses have also been made to arable fields.

Most traditional orchards seem to survive on the claylands in the south and centre of the county, where they were most numerous in the late nineteenth century: but wherever they survive, and even in degraded condition, they are of considerable historical as well as biological importance. As discussed in Chapter 2, apples and pears seldom attain any very considerable girth – rarely more than 2m, and never above 3m (see Figure 73). But they rapidly attain 'veteran' status. Oaklands Farm, just to the north of the town of Wymondham, had two orchards, one already in existence when the Ordnance Survey first edition 6-inch map was made in 1882, the other described in sales particulars from July 1891 as 'recently planted' (Wymondham Town Archives, uncatalogued). Four trees survive from the latter, almost certainly dating from the time of the original planting and thus around 120 years old. They have girths of only 0.85m, 1.0m, 1.1m and 1.2m respectively, yet already exhibit a range of classic 'veteran' features – cracks, cavities, broken and rotten wood. Two trees from the earlier orchard, to the south of the farmhouse, also remain, with slightly larger girths (1.65m and 1.3m respectively), but their age is uncertain. Such survivals – and, in spite of the scale of recent losses, there are still many in the county – need to be treated with the care and respect which they deserve. Every attempt should be made to restore and restock such remnant orchards, as well as to conserve those more completely preserved.

CHAPTER SEVEN

Conclusion

In this book we have considered the old and traditionally managed trees which are found in just one English county, Norfolk. We have brought together a number of separate surveys and pieces of research and drawn on the work and expertise of a number of different groups and individuals. Our analysis has not, as we have emphasised, been based on a *total* record of the old trees, or old trees of particular kinds, surviving in the county, and there are many other ancient and interesting trees and collections of trees yet to be discovered in the Norfolk countryside. Our conclusions are, however, based on the systematic mapping and analysis of a fairly substantial body of data, and many of them probably have a wider relevance to other parts of England. So too, more importantly, may our overall approach. The study we have described was carried out from a particular perspective – that of the landscape historian, rather than that of the natural historian or arboriculturalist. Practitioners of these latter disciplines have their own particular insights to offer, and it is not our intention to suggest that landscape history necessarily provides, in any sense, a better way of studying old or traditionally managed trees. It can, however, certainly make its own distinctive contribution to this fascinating subject. Our concern throughout, as should by now be apparent, has been to examine trees within their particular spatial and historical contexts: to seek meaningful patterns in the distribution of old trees, and of trees of particular kinds, and to explain these in terms of past social and economic, as much as environmental, processes. Such an historically and geographically focused approach throws new light on a number of important issues.

The kind of large-scale sampling and systematic mapping which has formed the basis for this research is currently continuing in a number of English counties, largely owing to the excellent work of the Tree Council. While it is not yet possible to see how Norfolk fits into the wider national (still less European) pattern, it is clear that different regions of the country boast very different legacies of ancient and traditionally managed trees: something which is hardly surprising, given England's diverse range of soils, climates and agricultural and social histories. Even a cursory perusal of the available evidence reveals, for example, that ancient pollards are a much more common feature of the countryside in counties like Hertfordshire – largely enclosed into hedged fields before the eighteenth century – than they are in Northamptonshire or Leicestershire, where large tracts of ground were occupied by unhedged open fields well into the post-medieval period; and that Norfolk, with both anciently enclosed and recently enclosed countryside, falls somewhere in between these extremes. At the same time a similarly superficial analysis reveals intriguing anomalies and raises important questions: why, for example, are old pollards noticeably less prominent in a county such as Shropshire, with an enclosure history broadly similar to that of the East Anglian counties? We have much to learn about the history of trees in the English landscape, and many of the arguments made in this book will doubtless need to be modified as research progresses elsewhere.

This said, some of our basic conclusions may stand the test of time. Some, indeed, are merely statements of what many others have already suggested or suspected, albeit sometimes on the basis of less evidence. In particular, it seems clear that while the distribution of ancient trees owes something to natural factors – soils, geology, hydrology and climate – human agency has played a fundamental role in structuring our arboricultural inheritance. Indeed, at the most basic level the kinds of trees found in the English countryside bear, at best, only a tangential relationship to those which made up the 'natural' vegetation of the country (whatever precise form that took) before humans began to make a significant impact on the environment as graziers and farmers. Some species did not thrive because they were adversely affected by farming and other activities: the almost complete disappearance of small-leafed lime from Norfolk and elsewhere, for example, is probably a consequence of the fact that it does not respond well to grazing pressure. It had probably declined even before the extensive wood-pastures of the pre-medieval period were largely cleared for agriculture or managed more intensively as coppiced woodland. More importantly, certain kinds of tree were encouraged, or deliberately planted, more than others over the centuries because they were of particular use to peasant farmers as wood or timber – such as oak and ash. Such practical considerations also largely determined how trees were managed, including, crucially, how long they were allowed to live before being felled. Even pollards would usually have been taken down in old age, as their productivity declined. This said, it also seems clear from the survival, well past economic maturity and long into senescence, of our very oldest specimens that trees might be regarded with real affection in the past and valued for social and ideological reasons as well as for practical and economic ones: the current interest in and affection for very old trees is clearly not new. People in the remote past were also drawn to them, aware that they were among the oldest features of the landscape and yet, like ourselves, living things. Large and ancient trees were regarded as symbols of family or community identity and displayed as objects of familial or communal pride. Particular specimens were valued as landmarks or boundary markers and were identified with particular figures, real or mythical – such as Robert Kett, one of Norfolk's great folk-heroes. Given the scale of the demand for wood and timber in the past, however, such considerations were seldom able to override more down-to-earth ones – indeed, it is a miracle that any tree has managed to survive into extreme old age, at least in an intensively settled county such as Norfolk. But the wealthier elements of society, in particular, were able to indulge an affection for old trees, and this explains the association which we have noted between our very oldest examples and the residences of the rich – and especially the landscape parks laid out around country houses in the course of the eighteenth and nineteenth centuries.

To emphasise such essentially human factors in structuring our legacy of ancient trees is not, of course, to deny the significance of environmental influences. The fact that the oldest trees in Norfolk are oaks reflects not only human choice but the particular characteristics of this species, which is long-lived and resilient; while, as we have seen, particular soil conditions and factors of water supply have helped to structure the distribution of veterans of particular species such as ash and hornbeam. Our legacy of ancient trees – both in terms of the kinds of old trees we have, and the places they are found – is, in short, the consequence of a complex interplay of 'human' and 'natural' factors. If our particular emphasis has been on the former, that is largely because others, more knowledgeable than us, have written at length on the latter.

Old trees thus need to be considered as part of the wider history of the landscape, and of the economic and agrarian systems that have shaped it over many centuries. But, as we hope we have demonstrated, trees – like many other features of the landscape – are in themselves a form of evidence about the past, in some circumstances just as important as the kinds of documents employed by conventional historians, but usually most powerful when used in combination with these. The concentrations of old pollards found hidden away in more recent woodland at a number of places in the county – and others doubtless await discovery – are clear evidence that wood-pastures continued to be managed in Norfolk well into the post-medieval period. The fact that many of these survivals are to be found on acid, sandy soils suggests – when combined with documentary evidence – that many areas of open heathland in Norfolk, and perhaps elsewhere, developed only in relatively recent times, rather than in remote prehistory, while the fact that most appear to be located on former common land suggests that communities in the past were better able to regulate their exploitation of communal resources than historians and historical ecologists have sometimes assumed. In a similar way, the fact that significant numbers of very small pollards, probably less than a century old, can be found in the Norfolk countryside clearly suggests that the decline of this form of management was slower than some historians believe, while the distribution of such trees indicates, fairly clearly, that its continuance was encouraged (like the maintenance of small domestic orchards in the county before the twentieth century) by particular social and tenurial circumstances, principally the survival of significant numbers of small yeoman farmers.

On a less positive note, the use that we can make of trees as historical evidence is sadly limited by our inability to date them accurately. Much has been written about how trees might be dated by measuring their circumference, but such work has mainly concentrated on the maximum size that trees of a particular species might reach at a given age. What we have emphasised instead is the great *variation* in growth rates – something which is widely recognised, but not perhaps sufficiently stressed, by most writers on old trees. Not only will trees of different species, or of the same species but growing in different conditions, put on girth at very different rates but, more importantly, trees of the same species growing in close proximity to each other will frequently exhibit very different dimensions after only a few decades. Such differences, once established, do not appear to lessen as growth rates decline in later life. We are thus unable to date individual trees with any accuracy or confidence, and the extent to which we can use trees as an independent historical source is thus to some extent limited although – as we hope we have shown – in many other ways a study of trees *en masse*, within their landscape contexts, can throw much light on both human and natural history.

Many writers have emphasised that trees do not have to be of any great age to be 'veterans', and thus of crucial importance in maintaining biodiversity. We would add that trees do not have to be very old to be historically interesting – like the young oak pollards just noted – or to make a major contribution to the character of individual landscapes, most notably in the case of the pine 'rows' of Breckland discussed in the previous chapter. Recent as much as early history needs to be examined if our legacy of old trees is to be fully appreciated and understood. Indeed, the particular balance of species found in any landscape – especially the kinds of hedgerow trees, which contribute so much to an area's distinctive character – may not always be of any great age, and while old trees can tell us much about the past, they can also mislead. We have seen how the present dominance of oak in Norfolk hedges initially appears, from the fact that the majority of ancient hedge trees are oak, to be a long-standing feature of the county's landscape. But we have also explained how this impression is in part a misleading one, the consequence of the wholesale loss of elm and the longevity of oak over ash. In some areas this situation is now changing, as ash – common as a shrub in most Norfolk hedges, whatever their age – is increasingly allowed to grow from sapling to timber tree in unmanaged hedges at the same time as the existing population of oaks grows old and declines.

For the landscape is always changing, and this includes the number and the character of the trees within it – whether we are thinking of their age range or of the particular balance of species present in any locality. Threats to the countryside's trees continue: more on balance still die than are replaced, and there is a current spate of epidemic diseases – such as acute oak decline and sudden oak death, and 'leaf miner' and canker in horse chestnut – which may be related to climate change. How the landscape develops in the future will be the result, in large measure, of the choices made by landowners and farmers, and of policy decisions made by national and local government and government agencies concerning the size of grants available for environmental enhancement and how these are targeted. And in the last resort these will be political decisions, influenced by the lobbying power of groups such as the Tree Council. But replanting the countryside is itself not enough. The character of that planting needs to be carefully considered, not merely to maintain biodiversity and sustain wildlife but also to perpetuate the particular character of each historic landscape. And to understand more fully the ways in which this varies from place to place, and which aspects of a local countryside are ephemeral and which more persistent, will require research which is historical and archaeological as much as botanical in its focus.

Bibliography

Alexander, K. (1999) 'The invertebrates of Britain's wood pastures', *British Wildlife* **11** (**1**), 108–17.

Allison, K. J. (1957) 'The sheep-corn husbandry of Norfolk in the sixteenth and seventeenth centuries', *Agricultural History Review* **5**, 12–30.

Amyot, T. E. (1874) 'The Winfarthing Oak', *Transactions of the Norfolk and Norwich Naturalists' Society* **2**, 12–18.

Amyot, T. E. (1894) 'The Winfarthing Oak', *Transactions of the Norfolk and Norwich Naturalists' Society* **6**, 116.

Armstrong, P. (1985) *The Changing Landscape: The History and Ecology of Man's Impact on the Face of East Anglia*, T. Dalton, Lavenham.

Bacon, K. (2000) 'Enclosure in East Norfolk', unpublished MA dissertation, Centre of East Anglian Studies, University of East Anglia.

Bacon, R. N. (1844) *The report on the agriculture of Norfolk, to which the prize was awarded by the Royal Agricultural Society of England*, Norwich.

Bailey, M. (1989) *A Marginal Economy? East Anglian Breckland in the Later Middle Ages*, Cambridge University Press, Cambridge.

Baird, W. and Tarrant, J. (1970) *Hedgerow Destruction in Norfolk, 1946–1970*, University of East Anglia, Norwich.

Bannister, N. (1996) *Woodland Archaeology in Surrey: Its Recognition and Management*, Surrey County Council, Kingston upon Thames.

Barnes, G. (2003) 'Woodland in Norfolk: A Landscape History', unpublished PhD thesis, School of History, University of East Anglia.

Barnes, G. and Skipper, K. (1995) 'Pollarded willows in the Norfolk Broads', *Quarterly Journal of Forestry* **89** (**3**), 196–200.

Barnes, G. and Williamson, T. (2006) *Hedgerow History: Ecology, History and Landscape Character*, Windgather Press, Macclesfield.

Barnes, G., Dallas, P., Thompson, H., Whyte, N. and Williamson, T. (2007) 'Heathland and wood pasture in Norfolk: ecology and landscape history', *British Wildlife* **18** (**6**), 395–403.

Barnes, G., Dallas, P. and Williamson, T. (2009) 'The black poplar in Norfolk', *Quarterly Journal of Forestry* **103** (**1**), 31–8.

Bevan-Jones, R. (2002) *The Ancient Yew: A History of* Taxus baccata, Windgather Press, Macclesfield.

Blatchley, J. ed. (1982) *D. E. Davy: A Journal of Excursions through the County of Suffolk 1823–1844*, Suffolk Records Society **24**, Woodbridge.

Blomefield, F. (1805) *An Essay Towards a Topographical History of the County of Norfolk*, 11 volumes, London.

Brasier, C. M. and Gibbs, J. M. (1973) 'Origin of the Dutch Elm Disease epidemic in Britain', *Nature* **242**, 607–9.

Briggs, P. A. (1998) 'Bats in trees', *Arboricultural Journal* **22**, 25–35.

Brown, D. (2001) 'Nathaniel Richmond (1724–84): "Gentleman Improver"', unpublished PhD thesis, Centre of East Anglian Studies, University of East Anglia.

Brown, J. (1861) *The Forester: A Practical Treatise on the Planting, Rearing, and General Management of Forest Trees*, Blackwood and Sons, Edinburgh and London.

Butcher, R. W. (1941) *The Land of Britain: Suffolk*, Geographical Publications, London.

Campbell, B. (1981) 'The extent and layout of commonfields in east Norfolk', *Norfolk Archaeology* **28**, 5–32.

Chambers, J. (1829) *A General History of the County of Norfolk*, London.

Clarke, W. G. (1908) 'Some Breckland characteristics', *Transactions of the Norfolk and Norwich Naturalists' Society* **8**, 555–78.

Clarke, W. G. (1925) *In Breckland Wilds*, London.

Clifford, S. and King, A. (2007) *The Apple Source Book: Particular Uses for Diverse Apples*, Hodder & Stoughton, London.

Coleman, N. (2002) 'Norfolk's Veteran Trees', unpublished MSc dissertation, University of Middlesex.

Collier, M. J. (2001) 'Thursford Wood Saproxylic Coleoptera Survey', unpublished survey for Norfolk Wildlife Trust.

Cook, M. (1676) *The Manner of Raising, Ordering, and Improving Forest-Trees*, London.

Cooper, F. (2006) *The Black Poplar: History, Ecology and Conservation*, Windgather Press, Macclesfield.

Cowell, F. (2009) *Richard Woods (1715–1793): Master of the Pleasure Garden*, Boydell and Brewer, Woodbridge.

Cushion, B. and Davison, A. (2003) *Earthworks of Norfolk*, East Anglian Archaeology **104**, Dereham.

Dallas, P. (2010a) 'Sustainable environments: common wood pastures in Norfolk', *Landscape History* **31** (**1**), 23–36.

Dallas, P. (2010b) 'Orchards in the Norfolk landscape: historic evidence of their management, contents and distribution', unpublished report for Norfolk County Council/Norfolk Biodiversity Partnership.

Daniels, S. (1988) 'The political iconography of woodland in the eighteenth century', in eds D. Cosgrove and S. Daniels, *The Iconography of Landscape*, Cambridge University Press, Cambridge, 51–72.

Darby, H. C. (1983) *The Changing Fenland*, Cambridge University Press, Cambridge.

Davison, A. (1973) 'The agrarian history of Hargham and Snetterton as recorded in the Buxton Mss', *Norfolk Archaeology* **35**, 335–55.

Davison, A. (1990) *The Evolution of Settlement in Three Parishes in South-East Norfolk*, East Anglian Archaeology **49**, Gressenhall.

Denne, M. P. (1987) 'Is sycamore native to Britain?', *Quarterly Journal of Forestry*, **81** (**2**), 201.

Dimbleby, G. W. (1962) *The Development of British Heathlands and their Soils*, Oxford University Press, Oxford.

Dolman, P. (1994) 'The use of soils disturbance in the management of grass heath for nature conservation', *Journal of Environmental Management* **44**, 123–40.

Dye, J. (1990) 'Change in the Norfolk Landscape: The Decline of the Deer Park', unpublished MA dissertation, Centre of East Anglian Studies, University of East Anglia.

East of England Apples and Orchards Project (2006) 'The condition of orchards in Norfolk', unpublished report for the Norfolk Biodiversity Partnership/Norfolk County Council. Accessible online at http://www.norfolkbiodiversity.org/reports%20and%20publications/default.asp.

Edlin, H. L. (1947) *Forestry and Woodland Life*, Batsford, London.

Edwards, M. (1986) 'Disturbance histories of four Snowdonia woodlands and their relation to Atlantic bryophyte distribution', *Biological Conservation* **37**, 301–20.

Elliott, B. (1986) *Victorian Gardens*, Batsford, London.

Evelyn, J. (1664; 2nd edn 1669) *Sylva: or a Discourse of Forest Trees*, 1st edn, London.

Eyre, S. R. (1955) 'The curving ploughland strip and its historical implications', *Agricultural History Review* **3**, 80–94.

Fay, N. (2002) 'Environmental arboriculture, tree ecology and veteran tree management', *Arboricultural Journal* **26**, 213–38.

Fleming, A. (1998) *Swaledale: Valley of the Wild River*, Edinburgh University Press, Edinburgh.

Freeman, J. ed. (1952) *Thomas Fuller: The Worthies of England*, George Allen and Unwin, London.

Gairdner, J. ed. (1895) *The Paston Letters, 1422–1509 A.D.*, Archibald Constable, Westminster.

Gilpin, W. (1809) *Observations On Several Parts of the Counties of Cambridge, Norfolk, Suffolk and Essex. Also On Several Parts of North Wales; Relative Chiefly to Picturesque Beauty, in Two Tours, the Former Made in the Year 1769. The Latter in the Year 1773*, London.

Green, T. (2001) 'Should ancient trees be designated as Sites of Special Scientific Interest?' *British Wildlife* **12** (**3**), 164–6.

Green, T. (2002) 'Arborists should have a central role in educating the public about veteran trees', *Arboricultural Journal* **26**, 239–48.

Green, T. (2010) 'The importance of open-grown trees', *British Wildlife* **21** (**5**), 334–8.

Gregory, J. (2005) 'Mapping improvement: reshaping rural landscapes in the eighteenth century', *Landscapes* **6** (**1**), 62–82.

Grigor, J. (1841) *The Eastern Arboretum, or Register of Remarkable Trees, Seats, Gardens &c in the County of Norfolk*, London.

Grubb, P. J., Green, H. E. and Merrifield, R. C. J. (1969) 'The ecology of chalk heath: its relevance to the calciole-calcifuge and soil acidification problems', *Journal of Ecology* **57**, 175–212.

Haggard, L. R. (1946) *Norfolk Notebook*, Faber, London.

Haggard, L. R. and Williamson, H. (1943) *Norfolk Life*, Faber, London.

Halstead, P. (1996) 'Ask the fellows who lop the hay: leaf fodder in the mountains of northwest Greece', *Rural History: Economy, Society, Culture* **9**, 211–35.

Harris, E. (1987) 'The case for sycamore', *Quarterly Journal of Forestry* **81** (**1**), 32–6.

Harris, L. E. (1953) *Vermuyden and the Fens: A Study of Cornelius Vermuyden and the Great Level*, Cleaver-Hume Press, London.

Hodder, K., Buckland, P., Kirby, K. and Bullock, J. (2009) 'Can the pre-Neolithic provide suitable models for re-wilding the landscape in Britain?' *British Wildlife* **20** (**5**) (special supplement), 4–15.

Hodge, C., Burton, R., Corbett, W., Evans, R. and Scale, R. (1984) *Soils and their Use in Eastern England*, Soil Survey of England and Wales, Harpenden.

Hogg, R. (1851) *British Pomology: Or the History, Description, Classification, and Synonyms of the Fruit Trees of Great Britain*, London.

Holderness, B. A. (1985) 'East Anglia and the Fens', in ed. J. Thirsk, *The Agrarian History of England and Wales Vol. 5, 1640–1750. 2, Agrarian Change*, Cambridge University Press, Cambridge, 119–245.

Howes, C. A. (2009) 'Ancient yew trees in the Doncaster landscape', *Arboricultural Journal* **32**, 91–6.

Jones, O. and Cloke, P. (2002) 'Non-human agencies: trees in place and time', in eds

C. Knappett and L. Malafouris, *Material Agency: Towards a Non-Anthropocentric Approach*, Springer, Guildford, 79–96.

Jessopp, A. (1887) *Arcady, for Better, for Worse*, London.

Keech, D. (2000) *The Common Ground Book of Orchards: Conservation, Culture and Community*, Common Ground, London.

Kent, N. (1796) *General View of the Agriculture of Norfolk*, London.

Ketton-Cremer, R. W. (1957) *Norfolk Assembly*, Faber, London.

Kirby, K. J. and Drake, C. M. eds (1993) *Dead Wood Matters*, English Nature, Peterborough.

Ladbrooke, R. (1843) *Views of the Churches of Norfolk*, 5 vols, Norwich.

Laird, M. (1992) *The Formal Garden: Traditions of Art and Nature*, Thames and Hudson, London.

Lambert, S. ed. (1975) *House of Commons Sessional Papers of the Eighteenth Century: George III; Reports of the Commissioners of Land Revenue*, Wilmington, Delaware.

Lawson, W. F. (1618) *A New Orchard and Garden*, London.

Lennon, B. (2009) 'Estimating the age of groups of trees in historic landscapes', *Arboricultural Journal* **32**, 167–88.

Liddiard, R. (2000) *'Landscapes of Lordship': Norman Castles and the Countryside in Medieval Norfolk 1066–1200*, BAR Brit. Ser. **309**, Oxford.

Liddiard, R. (2007) *The Medieval Park: New Perspectives*, Windgather Press, Macclesfield.

Long, S. H. (1933) 'The preservation of Kett's Oak', *Transactions of the Norfolk and Norwich Naturalists' Society* **13** (**4**), 356–60.

Loudon, J. C. (1838) *Arboretum et Fruticetum Britanicum, or, the Trees and Shrubs of Britain* …, London.

Loudon, J. C. (1843) *On the Laying Out, Planting and Managing of Cemeteries: And on the Improvement of Churchyards*, London.

Mabey, R. (1998) *Flora Britannica*, Chatto & Windus, London.

Markham, G. (1613) *The English Husbandman*, London.

Marshall, W. (1787) *The Rural Economy of Norfolk*, 2 vols, London.

Marshall, W. (1796) *Planting and Rural Ornament*, London.

Mitchell, A. (1974) *Field Guide to Trees of Britain and Northern Europe*, Collins, London.

Moreton, C. and Rutledge, P. eds (1997) *Skayman's Book: Farming and Gardening in Late Medieval Norfolk*, Norfolk Record Society **61**, Norwich.

Morgan, J. and Richards, A. (2002) *The New Book of Apples*, London.

Morris, J. (2003) 'Ancient woodland surveys in the Chilterns', in ed. M. Solick, *New Perspectives on Chiltern Landscapes*, Chiltern Conservation Board, Princes Risborough, 39–50.

Morton, A. (1998) *Tree Heritage of Britain and Ireland*, Swan Hill Press, London.

Mosby, J. E. G. (1938) *The Land of Britain: Norfolk*, Geographical Publications, London.

Muir, R. (2000) 'Pollards in Nidderdale: a landscape history', *Rural History* **11** (**1**), 95–111.

Muir, R. (2005) *Ancient Trees, Living Landscapes*, Tempus, Stroud.

Norden, J. (1618) *The Surveyor's Dialogue*, London.

Norfolk County Council (1975) *Farmland Tree Survey of Norfolk*, Norwich.

Oestmann, C. (1994) *Lordship and Community: The Le Strange family and the Village of Hunstanton, Norfolk in the First Half of the Sixteenth Century*, Boydell and Brewer, Woodbridge.

Orange, A. (1994) *Lichens on Trees*, National Museum of Wales, Cardiff.
Pakenham, T. (1996) *Meetings with Remarkable Trees*, Weidenfeld and Nicholson, London.
Parry, J. (2003) *Heathland*, The National Trust, London.
Peglar, S., Fritz, S. and Birks, H. (1989) 'Vegetation and land use history at Diss, Norfolk', *Journal of Ecology* **77**, 203–22.
Peterken, G. F. (2008) 'Wood-pasturage in the oakwoods', *Natur Cymru* Winter 2008, 4–7.
Petit, S. and Watkins, C. (2003) 'Pollarding trees: changing attitudes to a traditional land management practice in Britain, 1600–1900', *Rural History* **14** (**2**), 157–76.
Postgate, M. R. (1960) 'The Historical Geography of Breckland, 1600–1840', unpublished MA thesis, University of London.
Postgate, M. R. (1973) 'Field systems of East Anglia', in eds R. A. Butler and A. R. H. Baker, *Studies of Field Systems in the British Isles*, Cambridge University Press, Cambridge, 281–324.
Pryor, F. (2002) *Seahenge: A Quest for Life and Death in Bronze Age Britain*, Harper Collins, London.
Rackham, O. (1976) *Trees and Woodlands in the British Landscape*, Dent, London.
Rackham, O. (1980) *Ancient Woodland: Its History, Vegetation and Uses in England*, Arnold, London.
Rackham, O. (1986a) *The History of the Countryside*, Dent, London.
Rackham, O. (1986b) 'The ancient woods of Norfolk', *Transactions of the Norfolk and Norwich Naturalists' Society* **27** (**3**), 161–77.
Ravensdale, J. and Muir, R. (1984) *East Anglian Landscapes: Past and Present*, Joseph, London.
Read, H. J. (2000) *Veteran Trees: A Guide to Good Management*, English Nature, Peterborough.
Read, H. J. (2008) 'Pollards and pollarding in Europe', *British Wildlife* **19** (**4**), 250–59.
Read, H. J. and Frater, M. (1999) *Woodland Habitats*, Routledge, London.
Richens, R. H. (1983) *Elm*, Cambridge University Press, Cambridge.
Roberts, W. (1937) 'Richard Milles' new kitchen garden, 1765', *Journal of the Royal Horticultural Society* 62, 501–7.
Rodwell, J. S. (1991) *British Plant Communities Volume 2: Mires and Heaths*, Cambridge University Press, Cambridge.
Rogers, E. V. (1984) 'Sessile oak – *Quercus petraea* Leibe – in Norfolk', *Transactions of the Norfolk and Norwich Naturalists' Society* **17**, 291–7.
Rogers, E. V. (1993) 'The native black poplar (*Populus nigra* subsp *beautifolia*) in Norfolk', *Transactions of the Norfolk and Norwich Naturalists' Society* **29**, 375–82.
Rose, F. (1991) 'The importance of old trees including pollards for lichen bryophyte and epiphytes', in ed. H. J. Read, *Pollard and Veteran Tree Management*, Corporation of London, 28–9.
Rothera, S. prep. and ed. (1997) 'Breckland', available online at <http://www.english-nature.org.uk/science/natural/NA_Details.asp?NA_ID=46&S=&R=8>.
Rotherham, I. and Ardron, P. (2006) 'The archaeology of woodland landscapes: issues for managers based on the case-study of Sheffield, England and four thousand years of human impact', *Arboricultural Journal* **29**, 229–43.
Sanders, R. (2010) *The Apple Book*, Francis Lincoln, London.
Saunders, H. W. (1917) 'Estate management at Raynham 1661–86 and 1706', *Norfolk Archaeology* **19**, 39–67.
Silvester, R. (1999) 'Medieval reclamation of marsh and fen', in eds H. Cook and T.

Williamson, *Water Management in the English Landscape: Field, Marsh and Meadow*, Edinburgh University Press, Edinburgh, 122–40.

Skipper, K. (1989) 'Wood-Pasture: The Landscape of the Norfolk Claylands in the Early Modern Period', unpublished MA thesis, Centre of East Anglian Studies, University of East Anglia.

Slotte, H. (2001) 'Harvesting of leaf-hay shaped the Swedish landscape', *Landscape Ecology* **16**, 691–702.

Smith, A. H., Baker, G. M. and Kenny, R. W. (1982) *The Papers of Nathaniel Bacon of Stiffkey*, vol. 2, Centre of East Anglian Studies, Norwich.

Spray, M. (1981) 'Holly as fodder in England', *Agricultural History Review* **29**, 97–110.

Stamper, P. (2002) 'The tree that hid a king: the Boscobel Oak, Shropshire', *Landscapes* **3** (**1**), 19–34.

Stokes, J. and Hand, K. (2004) *The Hedge Tree Handbook*, The Tree Council, London.

Stokes, J. and Rodger, D. (2004) *The Heritage Trees of Britain and Northern Ireland*, Constable and Robinson and The Tree Council, London.

Strong, R. (1979) *The Renaissance Garden in England*, Thames and Hudson, London.

Sussams, K. (1996) *The Breckland Archaeological Survey 1994–1996. A Characterisation of the Archaeology and Historic Landscape of the Breckland Environmentally Sensitive Area*, Suffolk County Council, Norfolk Museums Service, English Heritage, Ipswich.

Taigel, A. and Williamson, T. (1991) 'Some early geometric gardens in Norfolk', *Journal of Garden History* **11** (**1–2**), 1–111.

Taylor, C. (1973) *The Cambridgeshire Landscape*, Hodder and Stoughton, London.

Taylor, C. (1999) 'Post-medieval drainage of marsh and fen', in eds H. Cook and T. Williamson, *Water Management in the English Landscape: Field, Marsh and Meadow*, Edinburgh University Press, Edinburgh, 141–56.

Thirsk, J. (1987) *England's Agricultural Regions and Agrarian History 1500–1750*, Routledge, London.

Thomas, K. (1983) *Man and the Natural World: Changing Attitudes in England 1500–1800*, Allen Lane, London.

Thomas, P. (2000) *Trees: Their Natural History*, Cambridge University Press, Cambridge.

Thomas, S. (2007) 'The Ancient Tree Hunt', *Quarterly Journal of Forestry* **101** (**3**), 223–6.

Thompson, F. M. L. (1963) *English Landed Society in the Nineteenth Century*, Routledge and Kegan, London.

Turner, M. (2005) 'Parliamentary enclosure', in eds T. Ashwin and A. Davison, *An Historical Atlas of Norfolk*, Phillimore, Chichester, 131–2.

Tusser, T. (1573) *Five Hundred Pointes of Good Husbandrie, as Well for the Champion or Open Countrie as Also for the Woodland, or Seuerall*, London.

Vera, F. (2002a) *Grazing Ecology and Forest History*, Cabi, Wallingford.

Vera, F. (2002b) 'The dynamic European forest', *Arboricultural Journal* **26**, 179–211.

Wade Martins, S. and Williamson, T. (1996) *The Farming Journal of Randall Burroughes 1794–1799*, Norfolk Record Society **58**, Norwich.

Wade Martins, S. and Williamson, T. (1999) *Roots of Change: Farming and the Landscape in East Anglia 1700–1870*, British Agricultural History Society, Exeter.

Wade Martins, S. and Williamson, T. (2008) *The Countryside of East Anglia: Changing Landscapes, 1870–1950*, Boydell and Brewer, Woodbridge.

White, J. (1997) 'What is a veteran tree and where are they all?' *Quarterly Journal of Forestry* **91** (**3**), 222–6.

White, J. (1998) 'Estimating the age of large and veteran trees in Britain', Forestry Commission Information Note 250, Alice Holt.

Whyte, N. (2008) *Inhabiting the Landscape: Place, Custom and Memory 1500–1800*, Windgather Press, Oxford.

Wilkinson, G. (1978) *Epitaph for the Elm*, Hutchinson, London.

Williamson, T. (1993) *The Origins of Norfolk*, Manchester University Press, Manchester.

Williamson, T. (1995) *Polite Landscapes: Gardens and Society in Eighteenth-Century England*, Alan Sutton, Stroud.

Williamson, T. (1996a) 'Roger North at Rougham: a lost house and its landscape', in eds C. Rawcliffe, R. Virgoe, and R. Wilson, *Counties and Communities: Essays on East Anglian History*, Centre of East Anglian Studies, Norwich, 275–90.

Williamson, T. (1996b) 'The planting of the park', in ed. A. Moore, *Houghton Hall: The Prime Minister, the Empress, and the Heritage*, Philip Wilson, London, 41–7.

Williamson, T. (1998) *The Archaeology of the Landscape Park: Garden Design in Norfolk, England, c.1680–1840*, BAR Brit. Ser. **268**, Oxford.

Williamson, T. (2000) *Suffolk's Gardens and Parks: Designed Landscapes from the Tudors to the Victorians*, Windgather Press, Macclesfield.

Williamson, T. (2002) *The Transformation of Rural England: Farming and the Landscape 1700–1870*, Exeter University Press, Exeter.

Williamson, T. (2006) *England's Landscape: Vol. 3, East Anglia*, English Heritage/Harper Collins, London.

Willis, P. (1977) *Charles Bridgeman and the English Landscape Garden*, Zwemmer, London.

Winchester, A. (1990) *Discovering Parish Boundaries*, Shire, Princes Risborough.

Yaxley, D. (2005) 'Medieval deer parks', in eds T. Ashwin and A. Davison, *An Historical Atlas of Norfolk*, Phillimore, Chichester, 56–7.

Yelling, J. A. (1977) *Common Field and Enclosure in England 1450–1850*, Macmillan, London.

Young, A. (1797) *General View of the Agriculture of Suffolk*, London.

Young, A. (1804) *General View of the Agriculture of Norfolk*, London.

Index

Entries in bold refer to the figures